Advances in Probability Distributions with Given Marginals

Mathematics and Its Applications

Managing Editor:

M. HAZEWINKEL

Centre for Mathematics and Computer Science, Amsterdam, The Netherlands

Volume 67

Advances in Probability Distributions with Given Marginals

Beyond the Copulas

Edited by

G. Dall'Aglio
Department of Statistics, Probability and Statistical Applications,
University "La Sapienza", Rome, Italy

S. Kotz
Department of Management Science and Statistics,
University of Maryland, College Park, U.S.A.

and

G. Salinetti
Department of Statistics, Probability and Statistical Applications,
University "La Sapienza", Rome, Italy

SPRINGER-SCIENCE+BUSINESS MEDIA, B.V.

Library of Congress Cataloging-in-Publication Data

Advances in probability distributions with given marginals : beyond
   the copulas / edited by G. Dall'Aglio, S. Kotz and G. Salinetti.
       p.   cm. -- (Mathematics and its applications ; v. 67)
   Lectures presented at a "Symposium on Distributions with Given
Marginals" organized by the Dept. of Statistics of the University La
Sapienza, Rome, Italy.
   Includes indexes.
   ISBN 978-0-7923-1156-0   ISBN 978-94-011-3466-8 (eBook)

   DOI 10.1007/978-94-011-3466-8

   1. Distribution (Probability theory)--Congresses.   I. Dall'Aglio,
Giorgio.   II. Kotz, Samuel.   III. Salinetti, G. (Gabriella), 1946-
.   IV. Università degli studi di Roma "La Sapienza." Dipartimento
di statistica, probabilità e statistiche applicate.   V. Title:
Marginals.   VI. Title: Copulas.   VII. Series: Mathematics and its
applications (Kluwer Academic Publishers) ; v.67.
QA273.6.A38   1991
519.2'4--dc20   91-2591

Printed on acid-free paper

SERIES EDITOR'S PREFACE

'Et moi, ..., si j'avait su comment en revenir, je n'y serais point allé.'

Jules Verne

The series is divergent; therefore we may be able to do something with it.

O. Heaviside

One service mathematics has rendered the human race. It has put common sense back where it belongs, on the topmost shelf next to the dusty canister labelled 'discarded nonsense'.

Eric T. Bell

Mathematics is a tool for thought. A highly necessary tool in a world where both feedback and nonlinearities abound. Similarly, all kinds of parts of mathematics serve as tools for other parts and for other sciences.

Applying a simple rewriting rule to the quote on the right above one finds such statements as: 'One service topology has rendered mathematical physics ...'; 'One service logic has rendered computer science ...'; 'One service category theory has rendered mathematics ...'. All arguably true. And all statements obtainable this way form part of the raison d'être of this series.

This series, *Mathematics and Its Applications*, started in 1977. Now that over one hundred volumes have appeared it seems opportune to reexamine its scope. At the time I wrote

"Growing specialization and diversification have brought a host of monographs and textbooks on increasingly specialized topics. However, the 'tree' of knowledge of mathematics and related fields does not grow only by putting forth new branches. It also happens, quite often in fact, that branches which were thought to be completely disparate are suddenly seen to be related. Further, the kind and level of sophistication of mathematics applied in various sciences has changed drastically in recent years: measure theory is used (non-trivially) in regional and theoretical economics; algebraic geometry interacts with physics; the Minkowsky lemma, coding theory and the structure of water meet one another in packing and covering theory; quantum fields, crystal defects and mathematical programming profit from homotopy theory; Lie algebras are relevant to filtering; and prediction and electrical engineering can use Stein spaces. And in addition to this there are such new emerging subdisciplines as 'experimental mathematics', 'CFD', 'completely integrable systems', 'chaos, synergetics and large-scale order', which are almost impossible to fit into the existing classification schemes. They draw upon widely different sections of mathematics."

By and large, all this still applies today. It is still true that at first sight mathematics seems rather fragmented and that to find, see, and exploit the deeper underlying interrelations more effort is needed and so are books that can help mathematicians and scientists do so. Accordingly MIA will continue to try to make such books available.

If anything, the description I gave in 1977 is now an understatement. To the examples of interaction areas one should add string theory where Riemann surfaces, algebraic geometry, modular functions, knots, quantum field theory, Kac-Moody algebras, monstrous moonshine (and more) all come together. And to the examples of things which can be usefully applied let me add the topic 'finite geometry'; a combination of words which sounds like it might not even exist, let alone be applicable. And yet it is being applied: to statistics via designs, to radar/sonar detection arrays (via finite projective planes), and to bus connections of VLSI chips (via difference sets). There seems to be no part of (so-called pure) mathematics that is not in immediate danger of being applied. And, accordingly, the applied mathematician needs to be aware of much more. Besides analysis and numerics, the traditional workhorses, he may need all kinds of combinatorics, algebra, probability, and so on.

In addition, the applied scientist needs to cope increasingly with the nonlinear world and the extra mathematical sophistication that this requires. For that is where the rewards are. Linear models are honest and a bit sad and depressing: proportional efforts and results. It is in the non-linear world that infinitesimal inputs may result in macroscopic outputs (or vice versa). To appreciate what I am hinting at: if electronics were linear we would have no fun with transistors and computers; we would have no TV; in fact you would not be reading these lines.

There is also no safety in ignoring such outlandish things as nonstandard analysis, superspace and anticommuting integration, p-adic and ultrametric space. All three have applications in both electrical engineering and physics. Once, complex numbers were equally outlandish, but they frequently proved the shortest path between 'real' results. Similarly, the first two topics named have already provided a number of 'wormhole' paths. There is no telling where all this is leading - fortunately.

Thus the original scope of the series, which for various (sound) reasons now comprises five sub-series: white (Japan), yellow (China), red (USSR), blue (Eastern Europe), and green (everything else), still applies. It has been enlarged a bit to include books treating of the tools from one subdiscipline which are used in others. Thus the series still aims at books dealing with:

- a central concept which plays an important role in several different mathematical and/or scientific specialization areas;
- new applications of the results and ideas from one area of scientific endeavour into another;
- influences which the results, problems and concepts of one field of enquiry have, and have had, on the development of another.

One of the perennial themes in mathematics is how to recover an entity from certain derived quantities. For instance, a function of complex variables from boundary values or a density function from all line integrals of it. There are many such problems and mostly they are far from easy. Just within probability and statistics there are already a host of such problems.

One class of such problems is concerned with how to determine a probability distribution from its marginal distributions. This has been the subject of serious investigations; the mathematics involved is deep and reaches far into various sophisticated domains and much has been discovered recently. The present volume, consisting of the invited contributions of a quite recent comprehensive meeting on the topic, gives a good picture of the current state of affairs.

The shortest path between two truths in the real domain passes through the complex domain.

J. Hadamard

La physique ne nous donne pas seulement l'occasion de résoudre des problèmes ... elle nous fait pressentir la solution.

H. Poincaré

Never lend books, for no one ever returns them; the only books I have in my library are books that other folk have lent me.

Anatole France

The function of an expert is not to be more right than other people, but to be wrong for more sophisticated reasons.

David Butler

Bussum, 5 March 1991

Michiel Hazewinkel

Table of Contents

PREFACE

The remarkable advances in the study of Distributions with given
marginals made necessary an occasion, for interested people, to meet,
exchange results, and discuss prospective developments in the field.
Thus a "Symposium on Distributions with Given Marginals (Fréchet
classes)" was organized by the Department of Statistics of the
University LA SAPIENZA , Rome, Italy, where important contributions to
the field were given during the first phase of development, in the
fifties. The meeting was dedicated to the memory of Giuseppe Pompily,
who had promoted these contributions. The Scientific Committee was made
by Giorgio Dall'Aglio (Univ. of Rome, Italy), Samuel Kotz (Univ. of
Maryland, USA) and Josef Štepán (Karlovy Univ.,Prague, Czechoslovakia).
Support by the University, the Consiglio Nazionale delle Ricerche, the
Istituto   Centrale di Statistica and the Istituto Nazionale delle
Assicurazioni helped the organization.

The timeliness of the initiative was confirmed by a large and
geografically distributed attendance (96 persons from 11 countries). The
stimulating climate of the meeting, the occasion to establish fruitful
research contacts, and the thorough information about both historical
developments and most recent results were largely appreciated by the
participants. A large collection of contributed papers enriched with
deeper details the subjects of the invited lectures.

The invited lectures here collected should give to the readers a
good acquaintance of the field. The list of contributed papers can help
to know people interested in the different subjects and to obtain
information through personal contacts.

As chairman of the Organizing Committee, I am glad to express
warm thanks to the lecturers, to the participants, and to the other
members of the Scientific and Organizing Committees, specially to
Dr. Elisabetta Bona and Mrs Adriana Cirenei.

Rome, January 1991.

Giorgio Dall'Aglio

FOREWORD

While preparing this volume for publication, the editors tried, to the best of their abilities, to eliminate as much as possible the common negative "side effects" associated with proceedings of a conference. A well-focussed, sigle-topic agenda of the Symposium allowed us to reduce substantially the lack of coordination between the papers appearing in this volume. Originally 12 papers were planned for inclusion. We thus sincerely regret that Professor C. Genest did not provide the text to the editors in due time. (Interested readers may wish to consult his important paper in *Biometrika* 74, 549-555 (1987) which contains some of the material presented at the Symposium).

The first two papers in this collection by G. Dall'Aglio (the initiator and the organizer of the Symposium, besides being a pioneer in the field of Fréchet classes), and B. Schweizer provide respectively a comprehensive survey of the subject matter of distributions with given marginals (also known as Fréchet classes) and the main tool for their study – copulas – whose development has been associated with Schweizer and A. Sklar's ground-breaking work which originated in the late '50s. The reader will, no doubt, appreciate the broad panorama of the subject matter described so vividly in these papers and the substantial advances that have been made at this juncture of pure mathematics and applied probability theory during the last 30 years. An addition of S. Bertino's paper in *Metron* 35, No. 1-2 (1977) may be appropriate to the list of references provided by Dall'Aglio. As to Schweizer's tour-de-force, note his discussion as of yet unpublished contributions by Darsow et al. (ref.8).

The next three papers by R.B. Nelson, M.J. Frank and the three authors P.Mikusinski, H. Sherwood and M.D. Taylor, are devoted to a more detailed and systematic study of copulas with an emphasis on statistical and distributional applications (Nelsen), probabilistic interpretations (Mikusinski et al.) and on algebraic aspects (Frank). Frank's fruitful algebraic ideas inspired Genest to construct families of useful and interesting probability distributions.

The next paper by S. Kotz and J. Seeger attempts – motivated by the notions of Fréchet bounds – to utilize the concept of copulas for constructive generation of multivariate distributions (densities) with a preassigned degree of dependence among the components. Examples of densities appearing in Nelsen's paper are quite relevant in this connection.

Kotz's and Seeger's contribution is also closely related to the next paper by M.H. Metry and A.R. Sampson in which the authors study the (partial) ordering – in terms of dependence – of bivariate distributions with given marginals. Interested readers may note that the *Proceedings of the Conference on Dependence* (of which Sampson was one of the organizers) due to appear in 1991 in the *Lecture Notes* pubblished by the

Institute of Mathematical Statistics, contain numerous papers on the topic in which Fréchet classes play and important role.

The next three papers by European contributors (L. Rüschendorf, H.G. Kellerer and the two authors V. Beneš and J. Štépán - the latter served as a member od the Scientific Committee of the Symposium) may be viewed as studies of measure-theoretic and topological generalizations of problems directly related to Fréchet classes and bounds. The authoritative and comprehensive survey by Rüschendorf (over 110 citations) is an invaluable reference source for the realm of problems connected with duality theorems for Fréchet bounds and applications to the theory of statistical inference. The third section of Rüschendorf's paper could be profitably contrasted to the Nelsen contribution appearing in this volume. Kellerer concentrates on duality theorems for marginal problems in a quite abstract setting - and provides an intriguing distinction between upper and lower semi-continuous cases which may find applications to statistical distributions. Beneš and Štépán survey extensively (also in an abstract setting) extremal solutions of the marginal problem. As their text and the useful list of references indicates, the work in this area, besides its relation to the Choquet theorem (a powerful tool for characterization of probability distributions), is also closely connected with the Mikusinski et al. contributions which deal with similar problem from a slightly different position.

The volume ends with a paper by S. Cambanis in which the "standard" multivariate distributions with given marginals - the well-known Eyraud-Farlie-Gumbel-Morgenstern family (EFGM) - is generalized. (This family plays a prominent role in Kotz and Seeger's paper). Cambanis analyses the difficulties associated with an extension of the EFMG family to an EFMG random process, thus opening a new and fruitful field for a future research.

In the published version of his inaugural lecture at the University of Leiden ("Missed chances", *CWI Quarterly* 2, 117-129, 1989) R.D. Gill convincingly demonstrates (based on three examples of research topics) that the prevalent notion as the distinction between pure and applied mathematics - the first being characterized by beautiful ideas, the second by possibly useful but dull calculations - is misplaced. It seems to the editors of this volume that the subject of distributions with given marginals is another example of the subtle interplay between applications and theory in mathematical sciences whose purpose is to "shift the limits of ability and knowledge as far as possible".

The editors apologize for the inevitable non-uniformity in the form and the style of the papers and for typos that might have inadvertently crept in. The diversity could perhaps be justified by the subject matter of this volume which deals with *distributions* for which *only the marginals* are fixed.

Samuel Kotz
December 1990

CONTRIBUTED PAPERS

ALSINA Claudi, Univ. Pol. Catalunya, SPAIN
On 2-positive semidefinite strict t-norm
BARR Aiala, Univ. of South Africa
Discrete bi-and multivariate distributions developed with help of the third marginal and ascending diagonal arrays
BEAM Kevin H., U.S. Military Academy
A new System of Multivariate distributions
BERANKOVA Petra, Univ. Karlovy, Prague
Extremal solutions of the general marginal problem
CUADRAS Carles M., Univ. of Barcelona
Distributions with given multivariate marginals and given intercorrelation matrix
DE LUCIA Luigi, Univ. of Rome
On the connection indices with special regard to normal double Distributions
HAVRANEK Tomas   Int.of Computer Sc., Prague
On statistical aspects of joint probability reconstruction
JIROUSEK Radim, Czech. Acad. Sciences
Iterative Proportional Fitting procedure and Decomposable Distributions
KOCH Giorgio, Univ. of Rome
On the mixture of Gaussian distributions with applications
MARSHAL Albert W., Univ. of British Columbia
Some connections between multivariate geometric and multivariate exponential distribution
MELIJSON Isaac, Tel Aviv Univ.
Sharp Bounds on the largest of some linear Combinations of Random variables with given marginal distributions
OLKIN Ingram, Stanford Univ.
Bivariate distributions generated from Polya Eggenberger urn models
QUESADA-MOLINA Jose Juan, Univ. de Granada
Copulas and multivariate dependence
RACHEV S.T. Zari, Univ. of California
Marginal Problems with Additional Constraints
REGAZZINI Eugenio, Univ. Bocconi, Milano
Positive dependence orderings: some historical remarks
SIMEONE Bruno, Univ. of Rome
On the maximization of the chi-square index in a Frechet Class
SKLAR Abe, Ill. Inst. Techn.
Copulas, and Marcov Processes
SMITH Cyril S., London School of Economics
On the bound for the L^2 Wasserstein Metric

SPIZZICHINO Fabio, Univ. of Rome
    **Families of distributions with marginals in given exponential
    classes and applications to filtering**
STOPPA Gabriele, Univ. of Trento
    **Alcune proprieta' delle distribuzioni proprie della classe di
    Gumbel multivariata**
STOYANOV Jordan, Bulgarian Acad. of Sciences
    **Properties of multidimensional distributions and their marginals:
    conterexamples and some open questions**
TIIT Ene-Margit, Tartu univ.
    **Construction of distributions with given marginals and given
    correlation matrix**
VENETOULIAS Achilles, M.I.T.
    Simulation of Families of Multivariate distributions
VITALE Rick, Univ. of Connecticut
    Approximation by MCD Processes
YANUSHEKEVICHIUS Roman, Lithuanian Acad. of Sciences
    **Factorization stability for multivariate Poisson distribution
    using stability of its marginals**

PARTICIPANTS

ALSINA Claudi	Univ. Politecnica Catalunya, Diagonal 649 08028 Barcelona, SPAIN
BARAGONA Roberto	Dip.Stat.Prob., Univ. La Sapienza, P.le A. Moro 5 00185 Roma ITALY
BARBIERI M.Maddalena	Dip.Stat.Prob., Univ. La Sapienza, P.le A. Moro 5 00185 Roma ITALY
BARR Aiala	Dpt. of Statistics, Univ. of South Africa P.O. Box 392, Pretoria, 0001 SOUTH AFRICA
BASSAN Bruno	Dip.Stat.Prob., Univ. La Sapienza, P.le A. Moro 5 00185 Roma, ITALY
BAYARRI M.J.	Dpt. de Estadistica & I.O., Univ. de Valencia, 46100 Burjasot, Valencia, SPAIN
BEAM Kevin M.	U.S. Military Academy, West Point, N.Y.12550,9999 , U.S.A.
BENES Viktor	Dpt. Math., Czech. Techn. Univ., Karlovo nam 13, 12135 Prague, CZECHOSLOVAKIA
BERANKOVA Petra	Fac. Math. and Physics, Karlovy Univ., Sokolovska 83, 18600 Prague 8, CZECHOSLOVAKIA
BERTI Patrizia	Dip. Statistico, Via Curtatone 1, 50123 Firenze, ITALY
BERTINO Salvatore	Dip.Stat.Prob., Univ. La Sapienza, P.le A. Moro 5 00185 Roma, ITALY
BOLOGNA Salvatore	Ist. di Matem. per la R.O., Fac. Econ. e Comm., Viale delle Scienze, 90128 Palermo, ITALY
BONA Elisabetta	Dip.Stat.Prob., Univ. La Sapienza, P.le A. Moro 5 00185 Roma, ITALY
BRUNI Carlo	Dip. di Informatica, Univ. La Sapienza, Via Eudossiana 18, 00184 Roma, ITALY
CAMBANIS Stamatis	Dpt. of Statistics, Univ. North Carolina, Chapell Hill, N.C. 27514, U.S.A.
CESARE Bianca Maria	Dip.Stat.Prob., Univ. La Sapienza, P.le A. Moro 5 00185 Roma, ITALY
CIFARELLI Michele	I.M.Q., Univ. Bocconi, Via Sarfatti 25, 20136 Milano, ITALY
COLOMBI Roberto	Univ. di Brescia, Corso Mameli 27, 25122 Brescia, ITALY
CRISMA Lucio	Dip. Mat. Appl., Univ. di Trieste, P.le Europa 1, 34100 Trieste, ITALY
CUADRAS Carles M.	Dpt. d'Estadistica. Univ. de Barcelona, Diagonal 645, 08028 Barcelona, SPAIN
DALL'AGLIO Giorgio	Dip.Stat.Prob., Univ. La Sapienza, P.le A. Moro 5 00185 Roma, ITALY

D'ARCANGELO Enzo   Dip.Stat.Prob., Univ. La Sapienza, P.le A. Moro 5
                           00185 Roma, ITALY

DE ANTONI Francesco   Fac. Econ.e Comm., Univ. di Pescara,
                           Viale Pindaro 42, Pescara, ITALY

DE LUCIA Luigi   Dip.Stat.Prob., Univ. La Sapienza, P.le A. Moro 5
                           00185 Roma, ITALY

FINOCCHIARO Maria   Dip.Stat.Prob., Univ. La Sapienza, P.le A. Moro 5
                           00185 Roma, ITALY

FRANK M.J.   Illinois Inst. of Techn., Chicago Ill.,
                           60616, U.S.A.

GALLO Francesca   Dip.Stat.Prob., Univ. La Sapienza, P.le A. Moro 5
                           00185 Roma, ITALY

GENEST Cristian   Dpt. of Math & Stat., Univ. Laval,
                           Cité Universitaire, Quebec G1K 7P4, CANADA

GIOVAGNOLI Alessandra Dip. di Matematica, Univ. di Perugia,
                           Via Vanvitelli 1, 06100 Perugia, ITALY

HAVRANEK Tomas   Inst. of Comp. Sci. CSAV, Pod Vodarenskou Vezi 2
                           18207 Prague, CZECHOSLOVACHIA

HERZEL Amato   Dip.Stat.Prob., Univ. La Sapienza, P.le A. Moro 5
                           00185 Roma, ITALY

HOLZER Silvano   Dip. di Mat. Appl., Univ. di Trieste,
                           P.le Europa 1, 34127 Trieste, ITALY

JIROUSEK Radim   Inst. of Inf. Theory & Autom.,
                           Pod Vodarenskou Vezi 4, 182-08 Prague 8,
                           CZECHOSLOVAKIA

KELLERER Hans   Math.Inst., Ludwig-Maximilians-Univ.,
                           Theresienstr. 39, D-8000 München 2, WEST GERMANY

KOCH Giorgio   Dip. Matem., Univ. La Sapienza, P.le A. Moro 5
                           00185 Roma, ITALY

KOTZ Samuel   College of Busin. & Manag., Univ. of Maryland,
                           College Park, Maryland 20742, U.S.A.

LETI Giuseppe   Dip.Stat.Prob., Univ. La Sapienza, P.le A. Moro 5
                           00185 Roma, ITALY

MAMMITZSCH Volker   Dpt. of Mathematics, Univ. of Marburg
                           Mans Merwein Str. Lahuberge, D-3550 MARGURG

MANFREDI Giuseppe   Dip.Stat.Prob., Univ. La Sapienza, P.le A. Moro 5
                           00185 Roma, ITALY

MANTOVAN Pietro   Lab. di Statistica, Univ. di Venezia,
                           30100 Venezia, ITALY

MARBACH Giorgio   Dip.Stat.Prob., Univ. La Sapienza, P.le A. Moro 5
                           00185 Roma, ITALY

MARSHALL Albert W.   Univ.of British Columbia, 2781 W. Shore Dr.Lummi
                           Island WA 98262, U.S.A.

MEILIJSON Isaac   Dpt. of Stat., Univ. of Tel Aviv, ISRAEL

MOLINARI Franco   Ist. di Informatica, Univ. di Trento,
                           Via Rosmini 42, 38100 Trento, ITALY

MORTERA Julia   Dip.Stat.Prob., Univ. La Sapienza, P.le A. Moro 5
                           00185 Roma, ITALY

NELSEN Roger B.   Lewis & Clark College, Portland, OR 97219, U.S.A.

OLKIN Ingram   Dpt. of Statistics, Stanford Univ.,
                           Stanford, CA 94305, U.S.A.

ORSINGHER Enzo   Dip. Stat. Prob., Univ. La Sapienza, P. le A. Moro 5
                             00185 Roma, ITALY
OTTAVIANI M. Gabriella   Dip. Stat. Prob., Univ. La Sapienza, P. le A. Moro 5
                             00185 Roma, ITALY
OTTAVIANI Riccardo   Dip. di Scienze Economiche, Univ. La Sapienza,
                             Via Nomentana 41, 00161 Roma, ITALY
PETRONE Sonia   I. M. Q., Univ. Bocconi, Via Sarfatti 25,
                             20136 Milano, ITALY
PICCINATO Ludovico   Dip. Stat. Prob., Univ. La Sapienza, P. le A. Moro 5
                             00185 Roma, ITALY
PIERACCINI Luciano   Fac. di Econ. e Comm., Univ. La Sapienza
                             Via Castro Laurenziano 9, 00161 Roma, ITALY
QUESADA-MOLINA Jose Juan   Dpt. de Mat. Aplicada, Universidad de Granada,
                             18071 Granada, SPAIN
RACHEV S. T. Zari   Dpt. of Statistics, University. of California,
                             Santa Barbara, CA 93106, U. S. A.
RACUGNO Walter   Dip. di Matematica, Viale Merello 49,
                             09100 Cagliari, ITALY
REGAZZINI Eugenio   I. M. Q., Univ. Bocconi, Via Sarfatti 25,
                             20136 Milan, ITALY
RICCIARDI Nicoletta   Dip. Stat. Prob., Univ. La Sapienza, P. le A. Moro 5
                             00185 Roma, ITALY
RIGO Pietro   Dip. Statistico, Universita' di Firenze,
                             50123 Firenze, ITALY
RIZZI Alfredo   Dip. Stat. Prob., Univ. La Sapienza, P. le A. Moro 5
                             00185 Roma, ITALY
ROCH Roy   Dpt. of Informat. & R. O., C. P. 6128 Succ. A
                             Montreal P. Q., CANADA H3C 3J7
RODRIGUEZ LALLENA José Antonio   Dpt. Mat. Aplicada, Univ. de Granada,
                             0471 Almeria, 18071 Granada, SPAIN
ROTONDI Renata   CNR, IAMI, Via A. M. Ampere 56, 20131 Milano,
                             ITALY
RUGGERI Fabrizio   CNR, IAMI, Via A. M. Ampere 56, 20131 Milano,
                             ITALY
RUIZ-RIVAS Carmen   Dpt. de Mat., Univ. Autonoma de Madrid,
                             28049 Madrid, SPAIN
RUNGGALDIER Wolgfang   Dip. di Matematica, Univ. di Padova,
                             Via Belzoni 7, 35131 Padova, ITALY
RUSCHENDORF L.   Inst. fur Math Stat., Westfalische Wilhelms-Univ.
                             Einsteistr. 62, 4400 Munster/WESTF.,
                             WEST GERMANY
SACCHETTI Dario   Dip. Stat. Prob., Univ. La Sapienza, P. le A. Moro 5
                             00185 Roma, ITALY
SALINETTI Gabriella   Dip. Stat. Prob., Univ. La Sapienza, P. le A. Moro 5
                             00185 Roma, ITALY
SAMPSON Alan   Dpt. of Mathematics & Statistics,
                             Univ. of Pittsburgh, Pittsburg PA 15260, U. S. A.
SAN MARTINI Dino   Dip. Stat. Prob., Univ. La Sapienza, P. le A. Moro 5
                             00185 Roma, ITALY
SCARSINI Marco   Fac. Econ. e Comm., Univ. La Sapienza
                             Viale Castro Laurenziano 9, 00161 Roma, ITALY

SCHWEIZER Berthold   Dpt. of Mathematics, Univ. of Massachusetts,
                         Amherst MA 01003, U.S.A
SEBASTIANI Paola   Dip. Stat. Prob., Univ. La Sapienza, P. le A. Moro 5
                         00185 Roma, ITALY
SEMPI Carlo   Dip. di Matematica, Univ. di Lecce, C.P. 193,
                         73100 Lecce, ITALY
SHERWOOD Howard   Dpt. of Mathematics, Univ. of Central Florida,
                         Orlando, Florida 32816-6990., U.S.A.
SHORACK Galen   Dpt. Statistic, Univ. Washington,
                         Seattle, WA 98195, U.S.A
SIGALOTTI Luciano   Dip. Mat. Appl., Univ. di Trieste, P. le Europa 1
                         34100 Trieste, ITALY
SIMEONE Bruno   Dip. Stat. Prob., Univ. La Sapienza, P. le A. Moro 5
                         00185 Roma, ITALY
SKLAR Abe   Dpt. of Mathematics, Ill. Inst. Tech.,
                         Chigago, ILL 60616, U.S.A.
SMITH Cyril S.   London School of Economics, Houghton St.
                         London WC2A 2AE, GREAT BRITAIN
SPEZZAFERRI Fulvio   Dip. Stat. Prob., Univ. La Sapienza, P. le A. Moro 5
                         00185 Roma, ITALY
SPIZZICHINO Fabio   Dip. Matematico, Univ. La Sapienza, P. le A. Moro 5
                         00185 Roma, ITALY
STEPAN Josef   Dpt of Statistics, Karlovy Univ., Sokolovska 83,
                         18600 Prague 8, CZECHOSLOVAKIA
STOPPA Gabriele   Ist. di Stat. & R.O., Univ. di Trento,
                         Via Verdi 26, 38100 Trento, ITALY
STOYANOV Jordan   Inst. of Mathematics, Bulgarian Acad. of Sciences
                         Box 373, 1090 Sofia, BULGARIA
TERZI Silvia   Dip. Stat. Prob., Univ. La Sapienza, P. le A. Moro 5
                         00185 Roma, ITALY
TIIT Ene Margit   Tartu University, J. Liivi 2, 202400 Tartu,
                         ESTONIA, URSS
TRANQUILLI G. Battista   Dip. Stat. Prob., Univ. La Sapienza, P. le A. Moro 5
                         00185 Roma, ITALY
VENETOULIAS Achilles   M. I.T., Sloan School, 1 Amherst Str., E40-133
                         Cambridge, MA 02139, U.S.A.
VERDINELLI Isabella   Dip. Stat. Prob., Univ. La Sapienza, P. le A. Moro 5
                         00185 Roma, ITALY
VIANELLI Luciana   Ist. di Matem. per la R.O., Fac. Econ. e Comm.,
                         Viale delle Scienze, 90128 Palermo, ITALY
VITALE Rik   Univ. Connecticut, Statistics U 120,
                         Storrs, CT 06269, U.S.A.
YANUSHEKEVICHIENE Olga   Inst. Math. Cibern., 4 Akademijos str., Vilnius
                         232600 LITUANIA, URSS
YANUSHEKEVICHIUS Roman   Inst. Math. Cibern., 4 Akademijos str., Vilnius
                         232600 LITUANIA, URSS
ZULIANI Alberto   Fac. di Econ. e Comm., Univ. La Sapienza
                         Viale Castro Laurenziano 9, 00161 Roma, ITALY

FRECHET CLASSES: THE BEGINNINGS

G. DALL'AGLIO
Department of Statistics
Piazzale A.Moro
I-00185 - Roma
Italy

ABSTRACT. After an introduction on the Symposium on Distributions with given marginals, the first phase in the development of the theory is shown. Preceeding appearances of the subject are recalled, especially those of the Roman statistical school and Hoeffding's work. The period considered is the decade starting with the initial paper by Fréchet in 1951. This paper examines the different directions in which researches developed, i.e. the structure of the class of distributions with given marginals, the distributions with given marginals and bounds, the compatibility of d.f.'s, the optimization of expectations and the related study of distance between distributions, the connection with the transportation problem of linear programming. This exposition ends with the introduction of the notion of copula by Abe Sklar.

1. Introduction

Advances in the field of probability distributions with given marginals have been remarkable in recent years. While it developed on its own, the problem was raised in several fields of application, so that the basic properties of Fréchet classes have been rediscovered more than once.

Thus it appeared necessary to create an occasion, for people interested in this subject, to meet, exchange results, discuss problems and make prospects on future developments.

I thought that Rome was an appropriate site for such a meeting. Even if the subject was not studied here in recent years, the location was justified by historical reasons. It was the Roman statistical school which made the first steps in this field; it was again in Rome that, after Maurice Fréchet laid the foundations, a large development took place. This development was due to Giuseppe Pompilj. The desire to pay homage to his memory, twenty years after his death, was yet another reason for promoting such a meeting here.

A confirmation of the partecipation of the Roman statistical school is given by the fact that both the present dean of the Facolta' di Statistica, Giuseppe Leti, and the director of the Department of Statistics, Alfredo Rizzi, have made contributions to the subject. I too

1

was involved; as a matter of fact distributions with given marginals have marked the first part of my scientific activity. This will help, I hope, to pardon the personal touch that I'm giving to this exposition.

*   *   *

Let us go back to our meeting. I sent a letter inquiring about people interested in a meeting on distributions with given marginals. I received a number of positive answers. To confirm the timeliness of the initiative, came a letter by Harold Sherwood with a similar proposal.

An enthusiastic answer was Sam Kotz's. He was an old friend. I had met him in the stimulating environment of the Department of Statistics at U.N.C., Chapel Hill, along with other remarkable people, such as H. Hotelling, W. Hoeffding, W. Smith, N. Johnson. I realize that I'm presenting a gallery of personnages; other will appear. Perhaps I should have given this talk a different title: "Frechet, Pompilj and I, not to mention many others".

2. Giuseppe Pompilj

Giuseppe Pompilj was an extraordinary man, rich in human and scientific qualities. He died in 1968, at the age of 54; in his not lengthy life he did a great deal for probability and statistics in Italy.

Pompilj was a geometrist, a student with Federico Enriquez and Guido Castelnuovo. His interest in probability and statistics came later, in an unusual way. During the war, he was a prisoner in India, and in the small library of the camp he found a book on mathematical statistics. In the same camp ther was also a student of this Facolta' who had to prepare his thesis; the help that Pompilj gave him was another factor which drew him to this field.

Thus Pompilj encountered mathematical statistics; a providential contact, which would have been difficult in his country. The Italian statistical school was then dominated by Corrado Gini, who was critical about the development that mathematical statistics was taking in Anglo-saxon countries, and drove the Italian school towards descriptive statistics. This direction was also favoured by the nationalism of the fascist period and by the separation induced by the war. So there was in Italy a deep ditch between statistics and probability, although the latter had famous scholars like Francesco Paolo Cantelli and Bruno de Finetti.

Back in Italy, Pompilj devoted himself to fill this gap, through teaching and promotion. At that time in this Facolta' probability was taught in the fourth and last year, clearly with no influence on the study of statistics. Pompilj taugth probability, enhancing it with sample theory, hypotesis testing, and design of experiments. Through his effort (not little, given the unelasticity of the Italian teaching system), probability found its natural allocation, at the second year, and with time mathematical statistics was spread in a number of subsequent courses. There was, in this changes, a substantial help by Gini, who recognized the importance of the subject in the formation of statisticians.

At the same time, Pompilj established frequent contacts with foreign countries, inviting lecturers and sending students abroad. He established a sort of "grand tour", in France and the United States. I was the first one to benefit from this organization.

Pompilj's role in distributions with given marginals is considerable. It is not manifest since, as in other fields, he preferred to let his students elaborate his ideas. So the work by Rizzi, Leti, Landenna, Bertino, Herzel and myself is in part due to him. An account of this work is contained in his book "Le variabili casuali" which appeared in 1984 after his death .

*   *   *

It was Pompilj who introduced me, indirectly, to distributions with given marginals. He asked Sandro Faedo, who then taught calculus at Pisa, to study a problem, and Faedo assigned it as the subject for my thesis. The problem dealt with was the minimization of

$$E|X-Y|^{\alpha} = \int_{R_2} |x - y|^{\alpha} \, d \, F(x,y) \tag{1}$$

when X and Y have given d.f. F_1 and F_2.

With Faedo's help I developed some results which, somewhat extended, appeared as a paper (1956) which later I had the satisfaction to see often quoted.

3. Maurice Frechet

In 1956, one year after graduation, I came to Rome, as an assistant to Professor Pompilj, continuing my researches with him.

Here I met Maurice Fréchet, who according to the foreign-oriented turn given by Pompilj, often visited our Istituto di Calcolo delle Probabilita'. Fréchet was then almost 80, but he didn't show it. After lectures and seminars we used to have dinner , and most Romans drove their cars, but he preferred to walk, and I often went with him, enjoyng his interesting conversation, ranging from mathematical observations to recent events. I remember discussing with him, in the spring of 1958, the power take-over by Charles De Gaulle in France, with illuminating remarks on a situation which I, as many in Italy, saw as too authoritarian.

I had of course frequent contacts with Fréchet in Paris, where I met other outstanding people, such as Daniel Dugué and Paul Lévy. Lévy also showed some interest in distributions with given marginals. In a note in 1960 he refers to a formula by Poincaré for utilization in n-dimensional distributions with given marginals, without developing the idea. However he played a role in the origins, as will be seen later.

Fréchet's famous paper which started the study of distributions with given marginals was published in 1951 in the Annales de l'Université de Lyon. The choice of a scarsely diffused journal was explained once by Fréchet with the desire to give more importance to "provincial" periodicals. This trait of his personality appeared also in his cultivation of Esperanto; his paper (1957c) has also a summary in this "helplanguage". I must add that I listened to a conversation in Esperanto between him and Jimmy Savage and drew the conclusion that they spoke two different languages, English Esperanto and French Esperanto.

*   *   *

Fréchet's approach to distributions with given marginals was through a paper by Paul Lévy, which appeared in Fréchet's 1950 book on probability. Lévy, searching for a definition of distance between two distributions, proves that, given a distance $d(X,Y)$ between random variables, the minimum of $d(X,Y)$, when the distribution of X and of Y are given, is again a distance; we will return to this again. It was probably through this paper that Fréchet got the idea of studying the class $\Gamma(F_1, F_2)$ of the bivariate d.f.'s F with given marginals F_1 and F_2.

He proved that

$$W(x,y) \leq F(x,y, \leq M(x,y)$$

were W and M are themselves d.f. of Γ, given by

$$W(x,y) = \max \{F_1(x) + F_2(y) - 1, 0\}$$

$$(2)$$

$$M(x,y) = \min \{F_1(x), F_2(y)\}$$

The extreme d.f.'s W and M are connected by a kind of duality. Consider r.v.'s -X ,Y , and denote by an apex the corresponding functions. We have

$$F'(x,y) = P(-X < x, Y < y) = F_2(y) - F(x+,y)$$

so that the d.f. W' corresponds to M , M' to W , and their properties exchange.

Fréchet also clarified the nature of the extreme d.f.'s W and M, showing that they represents joint distributions which are respectively "statistically decreasing" and "statistically increasing", i.e. they are concentrated on a non-increasing (non-decreasing) line.

Moreover Fréchet proved that the product joint d.f. is equal to one of the extreme ones if and only if at least one of F_1 and F_2 is degenerate, .so that Γ has only one element.

The problem, of course, was not new; we will see that in connection with the study of distance between distributions. Fréchet was the first to pose the problem in a systematic way. But the development of the subject was due to the convergence of other people's interest, which is mainly due to Giuseppe Pompilj.

* * *

Another question raised by Fréchet is that of the existence of distributions with given marginals and bounds, i.e. of d.f.'s F in Γ dominated by a given measure μ :

$$\int_B dF(u,v) \leq \mu(B) \qquad \text{for Borel sets B.}$$

For distributions concentrated on a finite number of values, he found the necessary condition

$$\sum_{i \in I} \sum_{j \in J} \mu_{i,j} - \sum_{i \in I} p_i - \sum_{j \in J} p'_j + 1 \geq 0 \tag{3}$$

for every subsets I and J of indices of the assumed values, were $\mu_{i,j}$ is the given measure of (x_i, y_j), and p_i, p'_j represents the given marginals. The sufficiency of (3) was proved by Fréchet for particular cases (1957a, 1958, 1960) and by Dall'Aglio in 1961. But a much greater generality was achieved independently in the same year 1961 by Hans Kellerer, who proved the sufficiency in R_2 of the generalization of (3), i.e.

$$\mu(A_1 \times A_2) - \int_{A_1} dF_1(x) - \int_{A_2} dF_2(x) + 1 \geq 0$$

for A_1, A_2 Borel sets, extending also the result to more general spaces.

* * *

A result worth mentioning is an average property, in the Fréchet class, of the independence two-way table, in a statistical setting (Leti, 1966). If $x_1, \ldots, x_n$ and y_1, y_n are two statistical series, and we take the two-way tables corresponding to all the permutations of the y indexes, for each case the arithmetic mean of all values is the independence term.

4. Distance between distributions

In the framework of distributions with given marginals, problem (1) can be formulated as the minimization of (1) when F belongs to $\Gamma(F_1, F_2)$.

The result is easily achieved through the relations, obtained by integration by parts,

$$E|X-Y| = \int_R [F_1(z) + F_2(z) - 2 F(z.z)] \, dz$$

and, for $\alpha > 1$,

$$E|X-Y|^\alpha = \alpha(\alpha - 1) \int_{u>v} [F_2(v) - F(u,v)](u-v)^{\alpha-2} \, dudv +$$

$$+ \alpha(\alpha-1) \int_{u<v} [F_1(u) - F(u,v)](v-u)^{\alpha-2} \, dudv \tag{4}$$

For $\alpha>1$ the unique minimizing d.f is $M(x,y)$. When $\alpha = 1$, there is a set of minimizing d.f.'s, all those which attain the maximum value on the line $y = x$. This set has obviously M as maximum element ; it also has a minimum (for any x and y) d.f., i.e.

$$F^*(x,y) = \begin{cases} F_1(x) - \max \{ \inf_{x\leq z\leq y} [F_1(z) - F_2(z)] , 0 \} & x \leq y \\[2mm] F_2(y) - \max \{ \inf_{y\leq z\leq x} [F_2(z) - F_1(z)] , 0 \} & x \geq y \end{cases} \tag{5}$$

A property of F^* is that it assignes to every subset of the diagonal $y = x$ the maximum probability compatible with marginals (Bertino, 1968a).

$$* \quad * \quad *$$

Optimization of (1) seems to be the starting point of all preceeding appearances of the idea of distributions with given marginals. The first one is probably due to the Roman statistical school.

In Corrado Gini's work the set of distributions with given marginals appears implicitly in several contexts. He criticized the coefficient of correlation on the basis that in general it cannot attain its extreme values 1 and -1 when the marginals are given. On another hand in 1914 he introduced the linear index of dissimilarity d given, for two statistical series $x_1, \ldots, x_n$ and $y_1, \ldots, y_n$, by

$$d = \frac{1}{n} \sum_1^n {}_r |x_r - y_r|$$

and justified this definition with the argument that it is the minimum of the distance

$$\frac{1}{n^2} \sum_1^n {}_{r,s} |x_r - y_s|$$

between two statistical variates, when the marginals are given; this minimum obtains when the variates are "cograduate". He follows the same procedure for the quadratic index

$$\left(\frac{1}{n} \sum_{1}^{n} {}_{r}(x_r - y_r)^2\right)^{\frac{1}{2}}$$

In other words Gini finds, in his finite setting, the minimum of (1) for $\alpha = 1, 2$. And in these definitions there is the same idea, that Lévy developed later, of distance between d.f. as the minimum of a distance between r.v.'s.

Many elaborations followed. A substantial step forward was made in 1939 by Tommaso Salvemini who constructed the "tabella di cograduazione" and "di controgradduazione" corresponding to the extreme d.f.'s of the Fréchet class.

The tabella di cograduazione was rediscovered as "N.W. corner rule" in the transportation problem of linear programming. It seems that this name was introduced in 1954 (see Leti, 1962)

The obvious connection with linear programming soon attracted attention and was initially investigated by Fréchet (1957b), then by Herzel (1961) and Féron (1963).

The dissimilarity index as a distance was often studied in that period. An interesting work is Landenna (1956), which considers the dissimilarity index in a general setting, exploiting the results on optimization of (1). The name of "Fréchet class" makes its appearence here.

*   *   *

In (1957b) Fréchet reexplores Lévy's definition of distance. He considers the class of distances between r.v.'s

$$d(X, Y) = f(|X-Y|) \tag{6}$$

were $f(z)$, for $z \geq 0$, is an increasing sub-additive function with $f(0)=0$, and according to the result for $f(z) = z^{\alpha}$, suggests that the distance between distributions can be defined as $d(X, Y)$ when the joint distribution of (X, Y) is M. By now, we know that this choice is not always consistent with the minimization principle: Bertino, (1968b) shows that the minimum in (6) obtains, under general conditions, for one of the distributions which minimize $E|X-Y|$; in particular for M when f is convex and for the d.f. F of (5) when f is concave.

*   *   *

Another forerunner was Wassilly Hoeffding, in 1940. He was the first to consider the set of distributions with given marginals, not only the extreme ones. He made even a reduction to the square of opposite vertices $(-1/2, -1/2)$ and $(1/2, 1/2)$, thus anticipating copulas. As for the optimization problem, he obtained for the covariance the expression

$$\sigma_{X,Y} = \int_R [F(u,v) - F_1(u)F_2(v)] \, dudv$$

(a neater form of (4) for $\alpha = 2$), thus finding the maximum and minimum values of the coefficient of correlation.

All the bases of the theory of distributions with given marginals are present here. But Hoeffding's paper appeared in Berlin at the beginning of World War II, on a scarcely diffused journal, and thus was neglected. His shy character did not help to made it known. I learned about the paper when I was in Chapel Hill, in his department, but not throug him.

5. Multivariate Frechet classes.

The need to go beyond two dimensions was soon realized. The most direct generalization is the class $\Gamma(F_1, F_2, \ldots, F_n)$ of n-dimensional d.f.'s whose one-dimensional marginals are given.

First results in this direction are in Félron (1956) for three dimensions, and in Rizzi (1957) for four dimensions. A generalization of the inequalities in two dimensions holds, with

$$W(x_1, \ldots, x_n) = \max \left\{ F_1(x_1) + \ldots + F_n(x_n) - n + 1, \, 0 \right\}$$

$$M(x_1, \ldots, x_n) = \min \left\{ F_1(x_1), \ldots, F_n(x_n) \right\}$$

The lack of symmetry between the two d.f. is evident, contrary to the two-dimensional case. In fact, while M is again a d.f., this is not always true for W, and the inequality for W cannot be improved, in the sense that in every point its value is attained by a d.f. of Γ. The reason for this lack of symmetry is clear if we look at the two-dimensional marginal d.f.'s

$$W^{(i,j)}(x_i, x_j) = \max \left\{ F_i(x_i) + F_j(x_j) - 1, \, 0 \right\}$$

which represent the maximum negative association: if both X_1 and X_2 have maximum negative association with X_3 then one should expect that between them there is a maximum positive association. So W is a d.f. only under very restrictive circumstances, with awkward effects. A necessary and sufficient condition, excluding the degenerate case, is that, setting

$$a_j = \inf \left\{ x : F_j(x) > 0 \right\} \qquad b_j = \sup \left\{ x : F_j(x) < 1 \right\}$$

it be either

$$F_1(a_1+) + \ldots + F_n(a_n+) \geq n - 1$$

or

$$F_1(b_1) + \ldots + F_n(b_n) \leq 1$$

In other words, all the distributions must start with a jump, and the sum of the jumps must be very high. The distribution under W is concentrated on n half lines parallel to the axes, starting from a common point, all positively or all negatively oriented (Dall'Aglio, (1960) for three dimensions, (1972) in general).

* * *

A Fréchet class can be studied also when the given marginals are not unidimensional. Consider a family $\mathcal{J}$ of subsets T of the set of integers $\{1,2,\ldots,n\}$. Given a set $\{ F_T(x_T); T\in\mathcal{J} \}$ of d.f.'s, one can study the class $\Gamma = \Gamma_{\mathcal{J}}$ of d.f.'s $F(x_1,\ldots,x_n)$ such that the marginal d.f.'s $F^T(x_T)$ are the given $F_T(x_T)$. It may be assumed that the union of sets T is equal to $\{1.2,\ldots,n\}$, otherwise the number of dimensions can be reduced.

If the sets T are pairwise disjoint, the product d.f. assures that the class Γ is not empty. But the properties of the class are rather elusive; it is not even assured that it possesses a maximum or minimum element (Rizzi, 1957).

* * *

When the sets T are not pairwise disjoint, the first problem is that of conditions for the compatibility of the d.f.'s F_T, i.e. for the existence of at least a d.f. belonging to the class Γ .

A necessary condition for compatibility can be immediately stated:

$$S, T \in \mathcal{J}, \qquad B = S \cap T \neq \emptyset \qquad \Rightarrow \qquad F_S^B(x_B) = F_T^B(x_B) \qquad (8)$$

i.e. the d.f.'s F_T must have the same marginals on common sub-spaces.

Condition (8) is not sufficient in general, but it is so with additional assumptions on the structure of family $\mathcal{J}$. Kellerer in 1964 characterized these assumptions.

Researches on compatibility, in the period we are examining, was limited to the case n = 3, focusing on the compatibility of the three d.f.'s

$$F_{1,2}(x_1, x_2), \qquad F_{1,3}(x_1, x_2), \qquad F_{2,3}(x_2, x_3)$$

The first result is by Jean Bass (1955), who gave a necessary and sufficient condition on the second moments (assumed finite). This direction of research was not developed, probably because at the time the interest in Fréchet classes was mainly in view of optimization of moments.

A different approach is followed in Dall'Aglio (1959), were the class of d.f.'s $F_{2,3}$ compatible with given $F_{1,2}$ and $F_{1,3}$ is studied; it is shown that this class is convex and has a maximum (in every point) element and a minimum one, giving thus the possibility of obtaining the extreme values of the moments. It is shown that, when $F_{1,2}$ and $F_{1,3}$ are both maximum or both minimum in their classes, under mild conditions (e.g continuity) the unique compatible $F_{2,3}$ is the maximum d.f. M.
This aspect was clarified later, as we have already seen.

The study of d.f's with given marginals dominated by a measure μ, which has already been examined, can be reduced to a special case of the compatibility problem. It suffices to consider the two-dimensional d.f. $F_{1,2}$ corresponding to $\mu/\mu(R_2)$ and a one-dimensional d.f. F_3 which assumes the values 0 and 1 with probability $1/\mu(R_2)$ and $1 - 1/\mu(R_2)$. If and only if these d.f.'s are compatible, there exists an $F(x_1, x_2, x_3)$ consistent with them, and the two dimensional d.f. $\mu(R_2)F(x_1, x_2, 1/2)$ is dominated by μ.

* * *

I've presented a panorama on the development of distributions with given marginals in the fifties, only mentioning some subsequent connected results.

In this decade a lot of good work had been made, exploring different lines of research: the structure of Fréchet classes, the distributions with given marginals and bounds, the compatibility of d.f.'s, the optimality of the mean value of some functions, the connection with linear programming. All these topics were to have a huge development, which for some of them started soon, as with Kellerer's work, alrealdy mentioned, which greatly improved the knowledge of the structure of the class, and with Strassen's results in stochastic processes. Most important was the introduction of copulas by Abe Sklar in 1961, and subsequent work by him and Schweizer, which opened a new way, bound to pervade all the field. But all that belonged to the future; the pioneer phase was concluded.

References

Bass, J. (1955) "Sur la compatibilité des fonctions de répartition", C. R. Acad. Sc. Paris 240, 839-841.

Benedetti, C. (1957) "Di alcune disuguaglianze collegate al campo di variazione di indici statistici", Metron 18 (fasc. 3-4), 102-125.

Bertino, S. (1968a) "Su di una sottoclasse della classe di Fréchet", Statistica 25, 511-542.

Bertino, S. (1968b) "Sulla distanza tra distribuzioni", Pubbl. Ist. Calcolo Probab. Univ. Roma 82.

Dall'Aglio, G. (1956) "Sugli estremi dei momenti delle funzioni di ripartizione doppia", Ann. Scuola Norm. Sup. Pisa, Cl. Sci (3) 10, 35-74

Dall'Aglio, G. (1959) "Sulla compatibilita' delle funzioni di ripartizione doppia", Rend. Mat. (5) 18, 385-413.

Dall'Aglio, G. (1960) " Les fonctions extremes de la classe de Fréchet a trois dimensions", Publ. Inst. Stat. Univ. Paris 9, 175-188.

Dall'Aglio, G. (1961) "Sulle distribuzioni con margini assegnati soggette a delle limitazioni", Giorn. Ist. Ital. Attuari 29, 94-108

Feron, R. (1956) "Sur les tableaux de corrélation dont les marges sont données, cas de l'éspace à trois dimensions", Publ. Inst. Stat. Univ. Paris 5, 3-12.

Feron, R. (1963) "Tableaux de corrélation dont les marges sont données et programmes linéaires", Publ. Inst. Stat. Univ. Paris 12, 103-116

Fréchet, M. (1951) "Sur les tableaux de corrélation dont les marges sont données", Ann. Univ. Lyon Sc. 4, 53-84.

Fréchet, M. (1957a) "Les tableaux de corrélation dont les marges et des bornes sont données", Ann. Univ. Lyon Sc. 20, 13-31.

Fréchet, M. (1957b) "Les tableaux de corrélation et les programmes linéaires", Revue Inst. Int. Stat. 25, 23-40

Fréchet, M. (1957c) "Sur la distance de deux lois de probabilité", Publ Inst. Stat. Univ. Paris 6, 185-189.

Fréchet, M. (1958) "Sur les tableaux de corrélation dont les marges et des bornes sont données", Revue Inst. Int. Stat. 28, 10-32.

Gini, C. (1914) "Di una misura della dissomiglianza tra due gruppi di quantita' e delle sue applicazioni allo studio delle relazioni statistiche", Atti R. Ist. Veneto Sc. Lett. Arti 74, 185-213.

Herzel, A. (1961) "Le tabelle di co- e contrograduazione e la programmazione linare", Metron 20, 186-200.

Hoeffding, W. (1940) "Massstabinvariante Korrelationtheorie", Schriften Math. Inst. Univ. Berlin 5, 181-233.

Kellerer, H. G. (1961) "Funktionen auf Producträumen mit vorgegeben Marginal Funktionen", Math. Annalen 153, 323-344.

Kellerer, H. G. (1964) "Verteilungfunktionen mit gegebenen Marginal-verteilungen" Z. Wahr 3, 247-270.

Landenna, G. (1956) "La dissomiglianza", Statistica 16, 21-57.

Leti, G. (1962) "Il termine generico delle tabelle di cograduazione e di contrograduazione", Biblioteca Metron C 1, 253-277.

Leti, G. (1966) "On the independence table for two distributions and the mean difference between them", Metron 25, 151-171.

Lévy, P. (1950) "Distance de deux variables aléatoires et distance de deux lois de probabilité", in Generalités sur les probabilités. Eléments aléatoires by M. Fréchet, Gauthier-Villars, Paris.

Lévy, P. (1960) "Sur la compatibilité des données marginales rélatives aux lois de probabilités", C. R. Acad: Sc. Paris 1250, 2507-2509.

Pompilj, G. (1984)   Le Variabili Casuali, Ist. Calcolo Probab. Univ.
    Roma.
Rizzi, A.   (1957) "Osservazioni sulle classi di Fréechet a piu'
    variabili", Boll. Unione Mat. Ital. 12, 263-277.
Salvemini, T. (1939)   "Sugli indici di omofilia", Proc. First Sc.
    Meeting Soc. Ital. Statist.; also Supplemento Statistica N.P. 5,
    105-115.
Sklar, A. (1959)   "Fonctions de répartition à n dimensions et leurs
    marges", Publ. Inst. Stat. Univ. Paris 8, 229-231.

THIRTY YEARS OF COPULAS[*]

BERTHOLD SCHWEIZER
Department of Mathematics and Statistics
University of Massachusetts
Amherst, Massachusetts 01003 USA

ABSTRACT. In 1959, in response to a query of M. Fréchet, A. Sklar introduced copulas. These are functions that link multivariate distributions to their one-dimensional margins. Thus, if H is an n-dimensional cumulative distribution function with one-dimensional margins $F_1, \ldots, F_n$, then there exists an n-dimensional copula C (which is unique when $F_1, \ldots, F_n$ are continuous) such that $H(x_1, \ldots, x_n) = C(F_1(x_1), \ldots, F_n(x_n))$. During the years 1959 - 1974, most results concerning copulas were obtained in the course of the development of the theory of probabilistic metric spaces, principally in connection with the study of families of binary operations on the space of probability distribution functions. Then it was discovered that two-dimensional copulas could be used to define nonparametric measures of dependence for pairs of random variables. In the ensuing years the copula concept was rediscovered on several occasions and these functions began to play an ever-more-important role in mathematical statistics, particularly in matters involving questions of dependence, fixed marginals and functions of random variables that are invariant under monotone transformations. Today, in view of the fact that they are the higher dimensional analogues of uniform distributions on the unit interval, and as the result of the efforts of a diverse group of scholars, the significance, ubiquity and utility of copulas is being recognized. This paper is devoted to an historical overview and rather personal account of these developments and to a description of some recent results.

---

[*]Research supported by ONR Contract N - 00014 - 90 - 1008.

0. INTRODUCTION

In July 1988, Professor G. Dall'Aglio wrote to me saying that he was
tentatively planning to organize a conference on distributions with given
marginals and asked whether "this was an initiative to be pursued".   I
responded with an enthusiastic "Yes"; and, presumptuously assuming that
I would be invited to speak, began to think about a possible topic.   An
historical survey seemed appropriate and the first title that came to
mind was "Copulas and their uses".   I soon realized that this was far too
ambitious an undertaking and therefore decided to limit my scope, to stay
with what I knew best, and to describe my personal involvement with the
subject over a span of three decades.   That was the substance of my talk
in Rome and that, in expanded form, is also the essence of this paper.

This paper is divided into two main parts.   The first part (Sections 1 - 5) is devoted to the years prior to the appearance of my joint
paper with E. F. Wolff on measures of dependence and describes the results that were obtained in conjunction with the development of the theory of probabilistic metric spaces.   The second part begins with a description of the joint work with Wolff (Section 6), continues with a -
necessarily incomplete - discussion of the relationship between our work
and the work of others (Section 7) and concludes with a summary of some
more recent developments (Sections 8 and 9).

In concluding this introduction I want to thank Professor Dall'Aglio for organizing the conference and for giving me and my colleagues
the opportunity to make our work known to a larger circle of scholars.
I also want to thank my colleagues A. Sklar, H. Sherwood, M. J. Frank
and R. B. Nelsen for the pleasure I have had working with them over the
years and for the help they have given me in the preparation of this
paper.

I. BEGINNINGS

A. Sklar and I began our collaboration on probabilistic metric
spaces in 1957.   By 1958 we had made some progress and submitted a note
describing our results to M. Fréchet for possible publication in the
Comptes Rendus de l'Académie des Sciences de Paris.   Fréchet accepted

our note [59] and with this an exchange of letters began. In one of
them Fréchet raised the question of determining the relationship between
a multidimensional probability distribution function and its lower dimensional margins - a question which had occupied him for a number of years
and to which he [20, 21], G. Dall'Aglio [4, 5] (see also [7]), R. Féron
[14] and others had made important contributions. In a subsequent letter to Fréchet, the substance of which later appeared in print [68],
Sklar answered this question for one-dimensional margins. He did so by
introducing the general notion of a copula.

DEFINITION 1.1. An *n-dimensional copula* (briefly, an *n-copula*) is
a function C from the unit n-cube $[0,1]^n$ to the unit interval $[0,1]$
which satisfies the following conditions:

(1.1) $C(1,\ldots,1,a_m,1,\ldots,1) = a_m$ for each $m \le n$ and all a_m
          in $[0,1]$.

(1.2) $C(a_1,\ldots,a_n) = 0$ if $a_m = 0$ for any $m \le n$.

(1.3) C is n-increasing

in the sense that the C-volume of any n-dimensional interval is non-
negative. In particular, if C is a 2-dimensional copula, then

(1.4) $C(a_2,b_2) - C(a_1,b_2) - C(a_2,b_1) + C(a_1,b_1) \ge 0$,

whenever $a_1 \le a_2$, $b_1 \le b_2$.

An n-copula may be viewed, or equivalently defined, as an n-dimensional cumulative probability distribution function whose support is
contained in $[0,1]^n$ and whose one-dimensional margins are uniform on
$[0,1]$.

It follows readily that any n-copula C is nondecreasing in each
argument separately and satisfies the Lipschitz condition

(1.5) $\left| C(b_1,\ldots,b_n) - C(a_1,\ldots,a_n) \right| \le \sum_{i=1}^{n} \left| b_i - a_i \right|$,

whence C is jointly continuous in all arguments and any collection of
n-copulas is equicontinuous. Furthermore, for any $n \ge 2$, the set of
all n-copulas is convex and is a compact metric space with respect to

the sup metric; indeed, in view of (1.5), in this space pointwise convergence and uniform convergence are equivalent. Lastly, the following inequalities hold:

(1.6) $\text{Max}(a_1 + a_2 + \ldots + a_n - n + 1, 0) \leq C(a_1, \ldots, a_n) \leq \text{Min}(a_1, \ldots, a_n)$,

for all $(a_1, \ldots, a_n)$ in $[0,1]^n$ and all $n \geq 2$. In particular, when n = 2,

(1.7) $\text{Max}(a + b - 1, 0) \leq C(a,b) \leq \text{Min}(a,b)$ for all (a,b) in $[0,1]^2$.

The lower bound, $\text{Max}(a + b - 1, 0)$ in (1.7) is a 2-copula; but as first
noted by Féron [14], for $n \geq 3$, $\text{Max}(a_1 + \ldots + a_n - n + 1, 0)$ is not an
n-copula; $\text{Min}(a_1, \ldots, a_n)$, on the other hand, is always a copula.

From (1.1) - (1.4) and (1.7) it follows that the graph of a 2-copula
is a continuous surface over the unit square that contains the skew quadrilateral whose vertices are (0,0,0), (1,0,0), (1,1,1) and (0,1,0).
This surface is bounded below by the two triangles that together make up
the surface of $z = \text{Max}(x + y - 1, 0)$ and above by the two triangles that
make up the surface $z = \text{Min}(x,y)$ (see Figure 1.1).

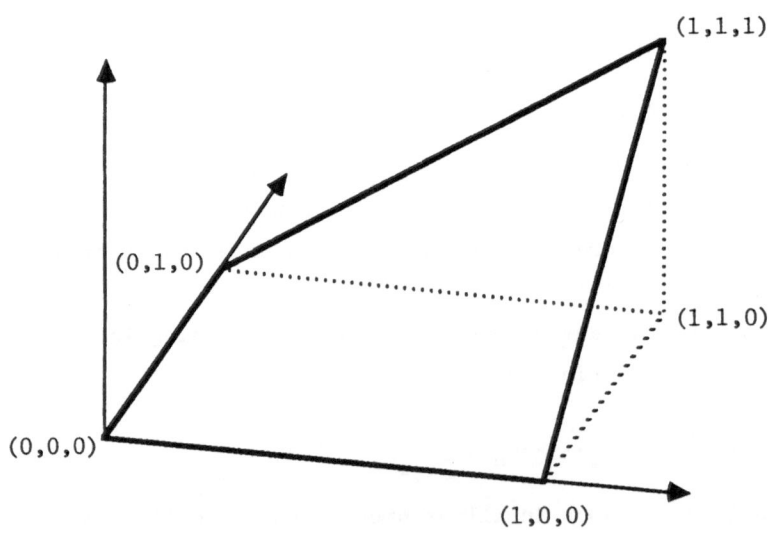

Figure 1.1.

Sklar's basic result is now the following:

THEOREM 1.1. If H is an n-dimensional probability distribution function with one-dimensional margins $F_1,\ldots,F_n$, then there exists an n-dimensional copula C such that, for all $x_1,\ldots,x_n$ in $\mathbf{R}$,

(1.8) $H(x_1,\ldots,x_n) = C(F_1(x_1),\ldots,F_n(x_n)).$

If H is continuous, then C is unique; otherwise C is uniquely determined on the Cartesian product $(\text{Ran } F_1) \times (\text{Ran } F_2) \times \ldots \times (\text{Ran } F_n)$. Conversely, if C is an n-dimensional copula and $F_1,\ldots,F_n$ are one-dimensional distribution functions, then the function H defined by (1.8) is an n-dimensional probability distribution function whose one-dimensional margins are $F_1,\ldots,F_n$.

Thus copulas are the functions that link n-dimensional distribution functions to their one-dimensional margins. Moreover, letting $u_i = F_i(x_i)$, $i = 1,\ldots,n$, yields

(1.9) $C(u_1,\ldots,u_n) = H(F_1^{-1}(u_1),\ldots,F_n^{-1}(u_n)),$

where, for any one-dimensional distribution function F, F^{-1} denotes the usual inverse of F which may, e.g., be defined by

(1.10) $F^{-1}(t) = \sup\{x \,|\, F(x) < t\}$.

This, together with (1.8), shows that much of the study of joint distribution functions can be reduced to the study of copulas. Note that (1.9) induces an equivalence relation on the space of n-dimensional distribution functions, where two such functions are equivalent (belong to the same Fréchet class) if and only if they determine the same copula.

When expressed in terms of random variables, the first part of Theorem 1.1 takes the following form:

THEOREM 1.2. Let $X_1,\ldots,X_n$ be real random variables, defined on a common probability space, with individual distribution functions $F_{X_1},\ldots,F_{X_n}$ and joint distribution function $H_{X_1 \ldots X_n}$. Then there exists an n-dimensional copula $C_{X_1 \ldots X_n}$ such that, for all $x_1,\ldots,x_n$ in $\mathbf{R}$,

$$(1.11) \qquad H_{X_1 \ldots X_n}(x_1, \ldots, x_n) = C_{X_1 \ldots X_n}(F_{X_1}(x_1), \ldots, F_{X_n}(x_n)).$$

If $F_{X_1}, \ldots, F_{X_n}$ are continuous, then $C_{X_1 \ldots X_n}$ is unique; otherwise $C_{X_1 \ldots X_n}$ is uniquely determined on $(\mathrm{Ran}\, F_{X_1}) \times \ldots \times (\mathrm{Ran}\, F_{X_1})$.

In the sequel, in order to avoid technicalities and to simplify the exposition (precise details may be found in the cited references), we shall generally assume that all random variables have continuous distribution functions. In this case we may refer to $C_{X_1 \ldots X_n}$ as the copula of $X_1, \ldots, X_n$. Also, since most of our discussion deals with, and most of our results apply to, pairs of random variables, we shall in the main confine ourselves to the two-dimensional case. We then have:

THEOREM 1.3. Let X and Y be random variables with continuous distribution functions F and G, respectively, joint distribution function H, and copula C. Then

(i)   X and Y are independent iff $C(a,b) = ab$.

(ii)   Y is a.s. an increasing function of X iff $C(a,b) = \mathrm{Min}(a,b)$.

(iii)   Y is a.s. a decreasing function of X iff $C(a,b) = \mathrm{Max}(a+b-1,0)$.

Moreover, on combining (1.7) and (1.8), we have

$$(1.12) \qquad \mathrm{Max}(F(x) + G(y) - 1, 0) \leq H(x,y) \leq \mathrm{Min}(F(x), G(y)),$$

and these inequalities hold generally, i.e., regardless of whether F and G are continuous or not.

Part (i) of Theorem 1.3 is of course well-known. Parts (ii) and (iii) - when expressed in terms of the joint distribution function H rather than the copula C - are generally attributed to Fréchet [20] although they are given (partly explicitly, partly implicitly) in the classic 1940 paper of W. Hoeffding [27]. The bounds in (1.12) are commonly known as the "Fréchet bounds" and the standard reference is Fréchet's groundbreaking 1951 paper [20]. Again, they may be found (in essence) in Hoeffding's paper. However, the attribution to Fréchet is entirely appropriate: for, as A. W. Marshall reminded me at the Symposium,

they already appear (for the n-dimensional case, as bounds on the probability of the joint occurrence of n events) in a paper which Fréchet
wrote in 1935 [19].

2. PROBABILISTIC METRIC SPACES

From 1958 to 1976 virtually all of the results concerning copulas
were obtained in connection with the study and development of the theory
of probabilistic metric spaces. In order to describe how this all came
about, we need to digress, define these spaces, and introduce the facts
that are relevant to our story. Details and further information may be
found in the book "Probabilistic Metric Spaces" written jointly with
A. Sklar [62].

Probabilistic metric spaces were first introduced by K. Menger in
1942 [44]. As defined in the 1958 Comptes Rendus note with Sklar, a
probabilistic metric space is an ordered pair (S, F), where S is a set
and F is a mapping from $S \times S$ into the space Δ^+ of probability distribution functions whose support is contained in the half-line $\mathbf{R}^+ = [0, \infty]$,
specifically,

(2.1) $\Delta^+ = \{F \mid \mathrm{Dom}\, F = [-\infty, \infty],\ \mathrm{Ran}\, F \subseteq [0,1],\ F$ is left-continuous
            on $(-\infty, \infty)$, nondecreasing, and $F(0) = 0,\ F(\infty) = 1\}$.

For any pair of points p, q in S the distribution function $F(p,q)$ is
generally denoted by F_{pq} and, for any real x, its value $F_{pq}(x)$ is usually interpreted as the probability that the "distance" between p and q
is less than x. The distribution functions F_{pq} are assumed to satisfy:

(I) $F_{pq}(x) = 1$ for all $x > 0$ iff $p = q$,

(II) $F_{pq}(x) = F_{qp}(x)$ for all x,

and a triangle inequality. The triangle inequality originally proposed
by Menger was

(2.2) $F_{pr}(u + v) \geq T(F_{pq}(u), F_{qr}(v))$,

for all points p, q, r in S and all $u, v \geq 0$, where T is a *t-norm*,
i.e., a mapping from $[0,1]^2$ into $[0,1]$ satisfying:

(2.2a) $T(a,1) = a$,

(2.2b) $T(a,b) = T(b,a)$,

(2.2c) $T(a,b) \leq T(c,d)$ whenever $a \leq c$, $b \leq d$,

(2.2d) $T(T(a,b),c) = T(a,T(b,c))$,

for all a,b,c,d in $[0,1]$.   It follows readily that

(2.3) $Z(a,b) \leq T(a,b) \leq Min(a,b)$,

where $Z(a,1) = Z(1,a) = a$ and $Z(a,b) = 0$ otherwise.   Thus the graph
of a t-norm is a surface over the unit square that contains the skew
quadrilateral whose vertices are $(0,0,0)$, $(1,0,0)$, $(1,1,1)$ and $(0,1,0)$
and which is bounded below by the graph of Z and above by the graph of
Min (see Figure 1.1).   Both Z and   Min are t-norms, as are the functions Prod and W defined on $[0,1]^2$ by

(2.4) $Prod(a,b) = ab$,

(2.5) $W(a,b) = Max(a + b - 1,0)$.

Note that   W, Prod and   Min   are copulas as well as   t-norms.

The conditions (2.2a), (2.2b) and (2.2c) are natural requirements
and have ready interpretations.   Conditions (2.2d), which was first introduced in [59], is needed in order to extend Menger's triangle inequality to a polygonal inequality.

Soon after the appearance of Menger's paper, A. Wald [75] proposed
an alternate triangle inequality, namely: For all p,q,r   in   S,

(2.6) $F_{pr} \geq F_{pq} * F_{qr}$,

where $*$ denotes convolution.   Subsequent developments have shown that
as a general requirement (2.6) is too stringent.

3.   t-NORMS AND COPULAS

The most interesting condition satisfied by a t-norm is the associativity condition (2.2d).   Accordingly, in the years 1961-63, Sklar and
I turned our attention to this functional equation [60].   We learned
that it has a long and distinguished history dating back to Abel and

became aware of the important results due to J. Aczel [1] (see also [2] and [62, Chapter 5]). These yielded the following:

THEOREM 3.1. Let T be a t-norm which is continuous on $[0,1]^2$ and strictly increasing on $(0,1]^2$. Then T admits the representation

(3.1) $T(a,b) = f^{-1}(f(a) + f(b))$ for all a,b in $[0,1]$,

where the one-place function f is continuous and strictly decreasing on $[0,1]$, with $f(1) = 0$ and $f(0) = \infty$, and f^{-1} is the inverse function of f. Conversely, given any such function f, the function T defined via (3.1) is a continuous t-norm. And there is a second function g such that $T(a,b) = g^{-1}(g(a) + g(b))$ if and only if $g = \lambda f$ for some $\lambda > 0$.

A t-norm satisfying the hypotheses of Theorem 3.1 is said to be *strict* and the function f in (3.1) is called an *additive generator* of T.

In [60] Sklar and I used the representation (3.1) to construct various one-parameter families of strict t-norms and thereby obtained a repertory of specific possible triangle inequalities of the form (2.2). In addition we also proved the following:

THEOREM 3.2. A strict t-norm is a 2-copula if and only if any, and hence every, additive generator is convex.

The above results were extended by C. H. Ling [41]. She showed that any continuous *Archimedean t-norm*, i.e., any continuous t-norm T satisfying

(3.2) $T(a,a) < a$ for all a in $(0,1)$,

admits the representation

(3.3) $T(a,b) = f^{(-1)}(f(a) + f(b))$ for all a,b in $[0,1]$,

where f is continuous and strictly decreasing on $[0,1]$ with $f(1) = 0$ and $f^{(-1)}$, the so-called *pseudo-inverse* of f, is defined via

(3.4) $f^{(-1)}(x) = \begin{cases} f^{-1}(x), & 0 \leq x \leq f(0), \\ \\ 0, & x \geq f(0). \end{cases}$

Theorem 3.2 remains valid when "strict" is replaced by "continuous Archimedean". Moreover, when combined with results of P.S. Mostert and A. L. Shields [46], it leads to a representation of any continuous t-norm as an "ordinal sum" (see [41] or [62]) of continuous Archimedean t-norms.

In the representation theorems of Aczel and Ling, the commutativity condition (2.2b) is a consequence of the other hypotheses; and this yields

THEOREM 3.3. A 2-copula is a t-norm if and only if it is associative.

Note also that among the t-norms W, Prod and Min, W is Archimedean but not strict, Prod is strict, and Min is not Archimedean.

We conclude this section with a brief mention of several related results that were obtained in subsequent years. The first is due to Sklar [69].

Let T be a t-norm. Then the *serial iterates* of T are the functions T^m, $m = 1,2,\ldots$, defined recursively via: Dom $T^m = [0,1]^{m+1}$, $T^1 = T$ and

$$(3.5) \qquad T^{m+1}(x_1,\ldots,x_{m+1},x_{m+2}) = T(T^m(x_1,\ldots,x_{m+1}),x_{m+2}).$$

Then the following holds:

THEOREM 3.4. Let T be a strict t-norm. Then the serial iterates of T are n-copulas for all $n \geq 2$ if and only if there is an additive generator f of T whose inverse g is completely monotonic, i.e., is real-analytic and satisfies $(-1)^n g^{(n)}(x) \geq 0$, for all x in $[0,\infty)$ and all $n \geq 0$. If one additive generator has a completely monotonic inverse, then all do. Furthermore, $T \geq$ Prod.

In [31, 32], C. H. Kimberling used Theorem 3.4 to characterize joint distribution functions of sets of exchangeable random variables.

Finally, we have the following result which is the companion of Theorem 3.3 and which is due to R. Moynihan [47]:

THEOREM 3.5. A t-norm T is a 2-copula if and only if it satisfies the Lipschitz condition

(3.6) $T(c,b) - T(a,b) \leq c - a$,

for all a,b,c in $[0,1]$ with $a \leq c$.

4. TRIANGLE FUNCTIONS AND σ-OPERATIONS

Menger's triangle inequality (2.2) involves a binary operation (t-norm) on the unit interval whereas Wald's triangle inequality (2.6) involves a binary operation (convolution) on the space of distribution functions Δ^+. In an important paper written in 1962, A. N. Šerstnev [65] (see also [66]) observed the following: The left-hand side of (2.2) depends only on the sum of u and v. Thus, letting $u+v=x$, it follows that (2.2) is equivalent to

(4.1) $F_{pr}(x) \geq \sup_{u+v=x} T(F_{pq}(u),F_{qr}(v))$,

for all x in **R**. Furthermore, it is not hard to show that if T is left-continuous in each place then, for any F, G in Δ^+, the function $\tau_T(F,G)$ defined on **R** via

(4.2) $\tau_T(F,G)(x) = \sup_{u+v=x} T(F(u),G(v))$

belongs to Δ^+. Thus τ_T, like convolution, is a binary operation on distribution functions. (Indeed, in convex analysis, such operations are frequently called "supremal convolutions".) Moreover, the operation τ_T satisfies conditions corresponding to (2.2a) - (2.2d). Accordingly, following Šerstnev, we define a *triangle function* to be a mapping τ from $\Delta^+ \times \Delta^+$ into Δ^+ satisfying:

(4.2a) $\tau(F,\varepsilon_0) = F$,

(4.2b) $\tau(F,G) = \tau(G,F)$,

(4.2c) $\tau(F,G) \leq \tau(H,K)$ whenever $F \leq H$, $G \leq K$,

(4.2d) $\tau(\tau(F,G),H) = \tau(F,\tau(G,H))$,

for any F,G,H,K in Δ^+. Here ε_0 is the unit step-function defined by

(4.3) $\varepsilon_0(x) = \begin{cases} 0, & x \leq 0, \\ \\ 1, & x > 0, \end{cases}$

and the ordering on Δ^+ is the usual pointwise partial ordering of functions. In addition, we will also assume that

(4.2e) τ is continuous in the (metrizable) topology of weak
          convergence of distribution functions.

   Examples of triangle functions are:

(i)   Convolution;

(ii)   The operations τ_T when T is a continuous t-norm;

(iii)   The operations π_T defined pointwise for any F,G in Δ^+ and
        any x in **R** via

(4.4) $\pi_T(F,G)(x) = T(F(x),G(x))$,

when T is a continuous t-norm; and several additional families which
will be introduced later.

   With the aid of the notion of a triangle function, we may now define a *probabilistic metric space* as an ordered triple (S,F,τ), where
S is a set, F is a mapping from $S \times S$ into Δ^+, τ is a triangle
function, and the following hold:

(I) $F(p,q) = \varepsilon_0$ if and only if p = q;

(II) $F(p,q) = F(q,p)$;

(III) $F(p,r) \geq \tau(F(p,q),F(q,r))$.

In short, probabilistic metric spaces are generalizations of ordinary
metric spaces in which $\mathbf{R}^+$, the range of the ordinary metric, is replaced by the set Δ^+, and the operation of addition on $\mathbf{R}^+$, which
plays a critical role in the triangle inequality, is replaced by a
triangle function, i.e., a well-behaved semigroup operation on Δ^+,
the range of the probabilistic metric.

   Coming back to copulas, the next step in the development stems
from a seminal observation of Z. Fiedorowicz: The operations τ_T
form a family of binary operations, not only on Δ^+, but also on the
space of probability distribution functions

(4.5) $D = \{F \mid \text{Dom}\, F = \mathbf{R},\ \text{Ran}\, F \subseteq [0,1],\quad F$ is left-continuous, non-

decreasing, $\lim_{x \to -\infty} F(x) = 0$, and $\lim_{x \to \infty} F(x) = 1\}$.

Fiedorowicz noted that convolution also belongs to such a family. To describe it, we first recall that for any distribution functions F and G we have

(4.6) $(F * G)(x) = \iint\limits_{u+v<x} d(F(u)G(v)) = \iint\limits_{u+v<x} d\,\text{Prod}(F(u),G(v)),$

and that if X and Y are independent random variables with individual distribution functions F_X and F_Y, then their joint distribution function is $\text{Prod}(F_X, F_Y)$ and the distribution function of their sum $X + Y$ is the convolution $F_X * F_Y$. Second, as is also well-known, if X and Y are arbitrary random variables with individual d.f.'s F_X and F_Y and joint d.f. H_{XY}, then the d.f. of their sum is given by

(4.7) $F_{X+Y}(x) = \iint\limits_{u+v<x} dH_{XY}(u,v).$

But by Sklar's Theorem,

(4.8) $H_{XY}(u,v) = C_{XY}(F_X(u), F_Y(v)),$

where C_{XY} is the copula of X and Y. Thus (4.7) may be rewritten in the form

(4.9) $F_{X+Y}(x) = \iint\limits_{u+v<x} dC_{XY}(F_X(u), F_Y(v)).$

And this, combined with (4.6), motivates the following:

DEFINITION 4.1. Let C be a 2-copula. Then σ_C is the binary operation on the space of distribution functions D defined via

(4.10) $\sigma_C(F,G)(x) = \iint\limits_{u+v<x} dC(F(u),G(v)).$

In particular, σ_{Prod} is convolution; and [17, 62] it can be shown that

(4.11) $\sigma_{\text{Min}} = \tau_{\text{Min}}.$

We can go further. If B is the class of Borel-measurable two-place functions, then for any L in B the d.f. of $L(X,Y)$ is given by (4.9) with $X + Y$ and $u + v$ replaced by $L(X,Y)$ and $L(u,v)$, respectively; and this leads to a larger, two-parameter, family of binary operations on D, namely the operations $\sigma_{C,L}$ defined via

(4.12) $\sigma_{C,L}(F,G)(x) = \displaystyle\iint_{L(u,v) < x} dC(F(u),G(v))$.

For example, if for any $\alpha > 0$, we let L_α be defined by $L_\alpha(u,v) = (u^\alpha + v^\alpha)^{1/\alpha}$, then the operations σ_{C,L_α} are the α-convolutions which have been extensively studied by K. Urbanik [73].

The class of binary operations τ_T can be similarly enlarged by replacing the operation of addition in (4.2) by a function L belonging to a specified subclass L of B to yield a two-parameter family of binary operations $\tau_{T,L}$ defined on D via

(4.13) $\tau_{T,L}(F,G)(x) = \sup_{L(u,v) = x} T(F(u),G(v))$.

When restricted to $\Delta^+ \cap D$, this leads to a larger class of triangle functions (see Chapter 7 of [62] for details).

Starting in the early 1970's, M. J. Frank undertook a detailed study of the operations σ_C [15]. From the point of view of probabilistic metric spaces, where we are interested in obtaining a supply of triangle functions, it is natural to ask: When is σ_C associative? Frank's complete and surprising answer is given by the following:

THEOREM 4.1. Let C be a 2-copula and let σ_C be defined by (4.10). Then σ_C is associative if and only if one of the following holds:

(a) $C = \text{Min}$;

(b) $C = \text{Prod}$;

(c) C is an ordinal sum all of whose components are either Prod or Min.

The σ_C is rarely associative; and the requirement that C be associative is necessary but far from sufficient. A further necessary condition for the associativity of σ_C is that $\bar{C}$, the *dual copula* of C, which is defined by

(4.14) $\bar{C}(a,b) = a + b - C(a,b)$,

is also associative. In [16], Frank showed that the only copulas C

having the property that both C and $\bar{C}$ are associative are Min, Prod, W and the family of copulas C_s defined for $0 < s < \infty$, $s \neq 1$, by

$$(4.15) \qquad C_s(a,b) = \log_s\left(1 + \frac{(s^a - 1)(s^b - 1)}{s - 1}\right)\Big/ \log s \ .$$

These copulas are generated by the functions

$$(4.16) \qquad f_s(a) = -\log\left(\frac{s^a - 1}{s - 1}\right) ,$$

and it is easy to show that

$$(4.17) \qquad
\begin{aligned}
&\lim_{s \to 0} C_s(a,b) = \text{Min}(a,b), \\
&\lim_{s \to 1} C_s(a,b) = ab, \\
&\lim_{s \to \infty} C_s(a,b) = W(a,b) = \text{Max}(a + b - 1, 0).
\end{aligned}$$

We will return to "Frank's family" of copulas in Section 7. The reader is also referred to Frank's contribution to this volume for further details.

5. DERIVABILITY AND BOUNDS

We conclude the first part of this paper with a discussion of several additional results that were obtained in the 1970's in the course of our work on probabilistic metric spaces.

It is well known that convolution of d.f.'s corresponds to addition of random variables, i.e., that if F and G are given d.f.'s, then there exist independent r.v.'s X and Y such that F is the d.f. of X (or, as we shall henceforth often write, $df(X) = F$), $df(Y) = G$ and $df(X + Y) = F * G$. Similarly, for any 2-copula C and any Borel-measurable two-place function L, there exist r.v.'s X and Y such that $df(X) = F$, $df(Y) = G$, C is the copula of X and Y, and $df(L(X,Y)) = \sigma_{C,L}$. Since the binary operations τ_T play a prominent role in the theory of probabilistic metric spaces, and since $\tau_{\text{Min}} = \sigma_{\text{Min}}$, it is natural to ask: What binary operations on random variables correspond to these operations? The answer is: Generally none. To make this more precise, we need the following:

DEFINITION 5.1. A binary operation Φ on $\mathcal{D}$ is *derivable from a function on random variables* if there exists a Borel-measurable two-place function V satisfying the following condition: For any d.f.'s F and G in $\mathcal{D}$, there exist r.v.'s X and Y, defined on a common probability space, such that

$$df(X) = F, \quad df(Y) = G \quad \text{and} \quad df(V(X,Y)) = \Phi(F,G).$$

Then, as Sklar and I showed in 1974 [61], we have:

THEOREM 5.1. Let T be any (left-continuous) t-norm other than Min. Then τ_T is not derivable from any function on random variables.

The same is generally true for the operations $\tau_{T,L}$ given by (4.13), as well as for the operations $\rho_{C,L}$ defined in (5.6) below.

In his paper [75], Wald introduced a notion of betweenness by postulating that, for distinct points p,q,r of a probabilistic metric space satisfying (2.6), q lies between p and r if and only if

(5.1) $F_{pr} = F_{pq} * F_{qr}$,

and he showed that this relation has all the properties of ordinary metric betweenness. Wald's idea extends to an arbitrary probabilistic metric space (S,F,τ) where q is *Wald-between* p and r if and only if

(5.2) $F_{pr} = \tau(F_{pq}, F_{qr})$.

But now the situation is more complicated since an arbitrary triangle function may not possess all the pleasant properties of convolution.

In the late 1970's, Moynihan and I undertook a detailed study of Wald-betweenness as well as other betweenness relations in probabilistic metric spaces [48]. In applying our results to a certain class of such spaces - the so-called E-spaces which are generated by mappings from a probability space into a metric space - we were led to the problem of determining all d.f.'s F and G in Δ^+ for which

(5.3) $\tau_W(F,G) = \sigma_C(F,G)$

for some given copula C. This problem was solved in a joint paper with Sklar [49]. There the three of us showed that the following basic inequality holds: For any 2-copula C and a large class of functions L in L,

(5.4) $\tau_{W,L} \leqq \tau_{C,L} \leqq \sigma_{C,L} \leqq \rho_{C,L} \leqq \rho_{W,L}$,

i.e., for all F and G in Δ^+,

(5.5) $\tau_{W,L}(F,G) \leqq \tau_{C,L}(F,G) \leqq \sigma_{C,L}(F,G) \leqq \rho_{C,L}(F,G) \leqq \rho_{W,L}(F,G)$,

where $\rho_{C,L}$ is the binary operation defined on Δ^+ via

(5.6) $\rho_{C,L}(F,G) = \inf_{L(u,v) = x} \bar{C}(F(u),G(v))$,

and $\bar{C}$ is the dual-copula of C given by (4.14) (see [49] and [62]
for the precise technical details). The inequality (5.4) itself is not
too difficult to establish and for L = Sum was already given by Sklar
in his survey paper of 1973 [69]. The problem is to determine when
equality holds: This is always the case when C = Min. Otherwise, generally speaking, equality holds at any single place in (5.5), and hence
in (5.3), if and only if either F or G is a unit step-function. In
[49] these results were established only on Δ^+, but it is not hard to
show that they also hold on $\mathcal{D}$.

6. MEASURES OF DEPENDENCE

The results presented in the first part of this paper were all obtained in connection with and as offshoots of problems arising in the
theory of probabilistic metric spaces. Those of us working on these
matters had no formal training in statistics. Thus we were only tan-
gentially aware of possible statistical applications. Moreover, with
the notable exception of Sklar's original paper [68], our results were
presented in a novel context and published in journals not generally
read by statisticians. Thus the statistical community took little note
of our work. (In retrospect, this had its advantages. It allowed us
to let the mathematical machinery develop at its own pace and to let
the ideas and techniques mature in an environment uninfected by the con-
tentious competition that seems to surround so many "hot" topics.)

The situation began to change in the Spring of 1974 while I was on
sabbatical leave in Italy. In the course of looking up some references
in the mathematical library of the University of Pavia, I came across
a paper by A. Rényi entitled "On measures of dependence" [52]. In this

paper Rényi presents a list of properties that "quantities which are
used to measure the strength of dependence between two random variables"
should satisfy. I had seen the paper before and had considered using
some of Rényi's conditions to define a probabilistic inner product. Looking at it again, it suddenly struck me that, using copulas, I could construct such Rényi-type measures of dependence at will.

Recall that the graph of a 2-copula C is a surface over the unit
square bounded above by the graph of Min and below by the graph of W
(see Figure 1.1); that the copula Min corresponds to monotone increasing dependence, that W corresponds to monotone decreasing dependence, and that Prod, whose graph lies symmetrically between the graphs
of Min and W, corresponds to independence. Hence, any measure of distance between the surfaces $z = C(x,y)$ and $z = xy$ will be a measure
of dependence between pairs of random variables whose copula is C. The
first measure that came to mind was the L_1-distance which, when suitably normalized, is given by

$$(6.1) \qquad \sigma(X,Y) = 12 \int_0^1 \int_0^1 |C_{XY}(u,v) - uv| \, du \, dv \, ,$$

for any pair of random variables X and Y. I worked out some of its
properties and showed that, in essence, it satisfied most of Rényi's
conditions.

On my return to Amherst, I discussed these matters with E.F. Wolff
and we began to work on them together. This led to a preliminary an-
nouncement published in 1976 [63], to Wolff's doctoral dissertation [79]
a year later, and to a joint paper which appeared in the Annals of Statistics in 1981 [64]. In that paper we established the following:

THEOREM 6.1. Let X and Y be continuous random variables with
copula C_{XY}. Then the quantity $\sigma(X,Y)$ satisfies the following conditions:

(A) $\sigma(X,Y)$ is well-defined.

(B) $\sigma(X,Y) = \sigma(Y,X)$.

(C) $0 \leq \sigma(X,Y) \leq 1$.

(D) $\sigma(X,Y) = 0$ iff X and Y are independent, i.e., iff $C_{XY} = $ Prod.

(E) $\sigma(X,Y) = 1$ iff each of X, Y is a.s. a strictly monotone function of the other, i.e., iff $C_{XY} = $ Min or $C_{XY} = $ W. If $C_{XY} = $ Min, this function is increasing; if $C_{XY} = $ W, it is decreasing.

(F) If f and g are strictly monotone a.s. on Ran X and Ran Y, respectively, then $\sigma(f(X),g(Y)) = \sigma(X,Y)$.

(G) If the joint d.f. of X and Y is bivariate normal, with correlation coefficient r, then $\sigma(X,Y)$ is a strictly increasing function of $|r|$, specifically,

$$\sigma(X,Y) = \frac{6}{\pi} \text{Arcsin} \frac{|r|}{2}.$$

(H) If (X,Y) and (X_n,Y_n), $n = 1,2,\ldots$, are pairs of continuous r.v.'s with joint d.f.'s H and H_n, respectively, and if the sequence $\{H_n\}$ converges weakly to H, then $\lim_{n \to \infty} \sigma(X_n,Y_n) = \sigma(X,Y)$.

It follows from property (F) that σ is a measure of monotone dependence, i.e., a rank statistic.

The properties (A) - (H) differ from Rényi's original conditions. Rényi did not require the continuity condition (H). He required (E) to hold if either $X = f(Y)$ or $Y = g(X)$ for some Borel-measurable functions f and g; and (F) to hold for all bijections from R into R. However, as already foreshadowed by Rényi himself and discussed in more detail in [63, 79], these requirements are too strong. Indeed, P. Mikusiński, H. Sherwood and M. D. Taylor have recently shown that, in the presence of (A) - (D) (resp., (A) - (E)), Rényi's version of (E) (resp., (F)) is incompatible with (H). This follows from the fact (see Theorem 3.2 of their contribution to this volume) that for any $\varepsilon > 0$ and any given r.v.'s X and Y, there exist r.v.'s X^* and Y^*, with $df(X^*) = df(X)$ and $df(Y^*) = df(Y)$, and a bijection f such that $Y^* = f(X^*)$ and

$$\sup_{u,v \in [0,1]} |C_{XY}(u,v) - C_{X^*Y^*}(u,v)| < \varepsilon/12,$$

whence $|\sigma(X,Y) - \sigma(X*,Y*)| < \varepsilon$. The important special case in which X
and Y are independent was established earlier by G. Kimeldorf and
A. R. Sampson [35] and, as they pointed out, implies that in practice
using only the joint d.f. of two r.v.'s one cannot distinguish statistical independence from complete functional dependence. Results very
closely related to those of Mikusiński, Sherwood and Taylor have recently been obtained by R. Vitale and the reader is referred to his paper
[74] for further discussion of this seemingly paradoxical state of affairs.

In [64] Wolff and I also proved the following:

THEOREM 6.2. Let X and Y be random variables with continuous
distribution functions F_X and F_Y, joint distribution function H_{XY}
and copula C_{XY}. Then the following hold:

(i)   If f and g are strictly increasing a.s. on Ran X and Ran Y,
        respectively, then

(6.2) $C_{f(X)g(Y)} = C_{XY}$.

(ii) C_{XY} is the (restriction to the unit square of the) joint d.f.
        of the probability transforms $F_X(X)$ and $F_Y(Y)$.

(iii)   If F and G are given continuous d.f.'s, then the r.v.'s
        $F^{-1}(F_X(X))$ and $G^{-1}(F_Y(Y))$ have d.f.'s F and G, respec-
        tively, and copula C_{XY}.

(iv)   If f and g are strictly decreasing a.s. on Ran X and Ran Y,
        respectively, then the copulas C_1, C_2, C_3 of the pairs (f(X),Y),
        (X,g(Y)), (f(X),g(Y)), respectively, are independent of the
        particular choices of f and g and are given by

          $C_1(u,v) = v - C_{XY}(1 - u,v)$,

          $C_2(u,v) = u - C_{XY}(u,1 - v)$,

          $C_3(u,v) = u + v - 1 + C_{XY}(1 - u,1 - v)$.

Wolff and I then went on to say that "... for us the true importance
of copulas lies in a combination of Sklar's Theorem and Theorem 6.2. For,

from the structure of (4.8) and the fact that under a.s. strictly increasing transformations of X and Y the copula is invariant while
the margins may be changed at will, it follows that it is precisely the
copula which captures those properties of the joint distribution which
are invariant under a.s. strictly increasing transformations. Hence
the study of rank statistics - insofar as it is the study of properties
invariant under such transformations - may be characterized as the
study of copulas and copula-invariant properties."

We further noted that the L_2 and L_∞ distances are given, respectively, by

$$(6.3) \qquad \sigma_2(X,Y) = (90 \int_0^1 \int_0^1 [C_{XY}(u,v) - uv]^2 du\ dv)^{1/2},$$

$$(6.4) \qquad \sigma_\infty(X,Y) = 4\ \sup_{u,v\ \in\ [0,1]} |C_{XY}(u,v) - uv|.$$

As measures of dependence, σ_2 satisfies conditions (A) - (H) of Theorem 6.1 and σ_∞ satisfies all but condition (E). Lastly, in terms
of copulas, the familiar Spearman's ρ and Kendall's τ are given,
respectively, by

$$(6.5) \qquad \rho(X,Y) = 12 \int_0^1 \int_0^1 (C_{XY}(u,v) - uv)du\ dv,$$

$$(6.6) \qquad \tau(X,Y) = 4 \int_0^1 \int_0^1 C_{XY}(u,v)dC_{XY}(u,v) - 1;$$

and, similarly, the medial correlation, or Blomqvist's q, is given by

$$(6.7) \qquad q(X,Y) = 4C_{XY}(1/2,1/2) - 1.$$

The definitions of the measures of dependence given by (6.1) and
(6.3) - (6.7) extend to higher dimensions. In [80], Wolff introduced
higher dimensional analogues of the modified Rényi conditions (A) - (H)
and studied the n-dimensional versions of σ, ρ, and τ in this light.
Many properties of these measures carry over readily, but there are
differences because the lower bound W^n (see (3.5)) in (1.6) is not
a copula and the graph of $Prod^n$ lies much closer to the graph of W^n
than to the graph of Min^n.

7. CONNECTIONS

While Wolff was working on his dissertation [79], and later in the course of writing our joint paper [64], we became aware of the extensive literature related, directly and indirectly, to copulas and the role copulas play in questions of dependence. It is too vast to survey here: and so, with due apologies to those whose contributions are not mentioned, I will confine myself to some remarks about those works that are closest to the matter at hand.

First and foremost, there is the classical 1940 paper of Hoeffding [27]. In it, to each two-dimensional probability density h, with cumulative d.f. H, he associates a "normalized sum-function" S_H. This is a d.f. whose support is contained in the unit square $[-1/2,1/2]^2$ and whose one-dimensional margins are uniform. The function S_H and the copula $C_H \cdot$ of H are simply related by

(7.1) $S_H(u,v) = C_H(u + 1/2, v + 1/2)$,

for any u,v in $[-1/2,1/2]$, and

(7.2) $C_H(a,b) = S_H(a - 1/2, b - 1/2)$,

for any a,b in $[0,1]$. Thus, had Hoeffding chosen the unit square $[0,1]^2$ instead of $[-1/2,1/2]^2$ for his normalization, he would have discovered copulas, and in all likelihood Sklar's theorem as well. Hoeffding's paper contains a wealth of important results. For example, he has Figure 1.1 and with it the Fréchet bounds; he gives the representation (6.5) for Spearman's ρ; he introduces the quantity σ_2 in (6.3) as a measure of dependence; he studies the normalized sum-function of the bivariate normal distribution; and invariance under increasing transformations (Masstabinvarianz) is the driving force throughout. Unfortunately, his paper appeared shortly after the outbreak of World War II and in a relatively obscure German journal. Thus, although it is often cited, I doubt that it has been widely read. It is still fresh today. It merits translation and republication; and the good news is that this is happening: Hoeffding's collected papers will soon be published.

Since 1959, copulas have been rediscovered by several authors. The
first to do so were Kimeldorf and Sampson. In a paper published in 1975
[34], they used (1.9) to define two-dimensional copulas. They called
them *uniform representations;* and in [35] and several subsequent papers
[36, 37] they developed many of their basic properties and used them
as a tool to define and study various dependence notions. Further details and additional references may be found in Sampson's contribution
to this volume.

Copulas (n-dimensional) appear briefly as *dependence functions* in
the book [22] by J. Galambos. And in 1978, P. Deheuvels took up their
study. In [9], using (1.8) as a starting point, he gave a rigorous
definition of n-dimensional dependence functions, derived their salient
properties and gave an independent proof of Theorem 1.1. (Although
Sklar's proof of Theorem 1.1 for the particular case $n = 2$ had appeared in 1974 [61], his proof for the n-dimensional case remained unpublished until 1983 [62]. A third independent proof appears in a
paper published in 1975 by D. S. Moore and M. C. Spruill [45], where,
however, the function C in (1.8) remains anonymous.) Most of Deheu-
vel's paper [9] is devoted to a detailed study of the possible limit
distributions of the extreme values of an infinite sample from a multivariate distribution. Specifically, let $\{\bar{X}_n\}$ be a sequence of
i.i.d. random vectors in $\mathbf{R}^m$ with common m-dimensional copula C.
Let $\{\bar{Y}_n\}$ be the sequence of vectors in $\mathbf{R}^m$ whose k'th component is
given by $\sup_{1 \leq i \leq n} X_{ki}$, where for $k = 1,\ldots,m$, X_{ki} is the k'th
component of $\bar{X}_i$, and let C_n be the copula of $\bar{Y}_n$. Then (see [22,
p. 251]).

(7.3) $C_n(u_1,\ldots,u_m) = C^n(u_1^{1/n},\ldots,u_m^{1/n}).$

The limit points of sequences such as $\{C_n\}$ are *extreme dependence*
functions, i.e., copulas of extreme value distributions. Deheuvels
gave an integral representation for such extreme dependence functions
and also characterized the set of these functions. (See also the
paper [26] by C. Genest and L. P. Rivest.) Continuing, in [10] he defined the order of an m-dimensional copula C to be the least upper

bound of all real numbers r such that the function $C^r(u_1^{1/r}, \ldots, u_m^{1/r})$ is a copula and studied the class of copulas of infinite order, which includes the class of extreme value copulas. He also made the point that "dependence functions are the real key to the study of multivariate distributions". Next, in a series of papers beginning with [11] and including [12], where further references may be found, Deheuvels defined and undertook an in-depth study of the empirical copula of a sample. He investigated its properties as an estimator of the population copula, used it to construct a variety of distribution-free tests of independence as well as to obtain nonparametric estimates of extreme value distributions, and found exact and asymptotic distributions of various estimators, etc. . Lastly, in [13] he showed how the Fréchet bounds for an n-copula can be improved when one has additional information concerning the marginals, e.g., when one knows that some of them are mutually independent.

Of the many other pre-1980 papers that could be listed here, there is first of all the one by Dall'Aglio [6] in which he studied the relationship between convergence in distribution and convergence in probability and proved, among other things, the following:

THEOREM 7.1. A sequence of r.v.'s $\{X_n\}$ converges in probability to a r.v. X if and only if $\{X_n\}$ converges in distribution to X and the sequence of copulas $\{C_{X_n X}\}$ converges to Min.

Next there are the well-known papers by W. H. Kruskal [39] and E. L. Lehmann [40]. In terms of copulas, the measures of association to which Kruskal devotes most of his attention are given by (6.5), (6.6) and (6.7); and Lehmann's concept of positive quadrant dependence may be succinctly expressed by saying that the r.v.'s X and Y are *positively quadrant dependent* whenever

(7.4) $C_{XY} \geq \text{Prod}.$

Further discussion of these and related concepts of association and dependence, as well as further references to the literature, are given in R. B. Nelsen's and A. R. Sampson's contributions to this volume. Note

also that, in this context, if X and Y are r.v's whose copula satisfies the hypotheses of Theorem 3.4, then X and Y are positively
quadrant dependent.

There is another topic which is too extensive to deal with here,
namely the question of the compatibility of higher dimensional d.f.'s
or copulas. This was first studied by Dall'Aglio [4, 7] and Sklar
(see Section 6.6 of [62]), then from a measure theoretic point of view
by H. G. Kellerer (see, e.g. [29, 30]) and V. Strassen [70], and more
recently by L. Rüschendorf (see, e.g. [54]). The reader is referred to
Kellerer's and Rüschendorf's contributions to this volume for a discussion of various current aspects of these issues and for additional
references to the literature.

After the appearance of [64] a number of other people began to take
an interest in copulas. The first among them were S. Kotz and N.L. Johnson. In [38] they computed the measure of dependence $\sigma(X,Y)$ for several families of iterated generalized Farlie–Gumbel–Morgenstern distributions. They found that for these distributions the maximum value of
$\sigma(X,Y)$ is considerably less than 1, whence it follows that these distributions cannot be used to model strong dependence. More details are
given in S. Kotz's contribution to this volume.

In [55], M. Scarsini defined a *measure of concordance* for continuous r.v.'s X and Y to be a function $I(X,Y)$ that satisfies the
conditions (A), (B) and (H) of Theorem 6.1, as well as

(a) $-1 \leq I(X,Y) \leq 1$,

(b) $I(X,Y) = 0$ whenever X and Y are independent,

(c) $I(-X,Y) = -I(X,Y)$,

(d) $I(X,Y) \geq (I(W,Z)$ whenever $C_{XY} \geq C_{WZ}$.

He showed that if f is a bounded, monotone odd function then the function

(7.5) $I_f(X,Y) = k \int_0^1 \int_0^1 f(u - 1/2) f(v - 1/2) dC_{XY}(u,v),$

where $k = \int_0^1\int_0^1 f^2(u - 1/2)du$, is a measure of concordance. For $f(u) = u$

it is Spearman's ρ and for $f(u) = \text{sgn}\,u$ it is Blomqvist's q. Kendall's τ and Gini's G, which is given by

(7.6) $G(X,Y) = 2\int_0^1\int_0^1 (|1 - u - v| - |u - v|)dC_{XY}(u,v)$,

are also measures of concordance, but not of the form (7.4). Scarsini also suggested a way of using copulas (which amounts to using the empirical copula) to define measures of concordance in the discrete case. Continuing his studies, in [56] he introduced strong measures of concordance, for which the inequalities in (d) are strict; in [57] he used copulas to study multivariate stochastic dominance; and in [58] he showed that under certain conditions a representation analogous to (1.8) can be obtained for any probability measure on a Polish product space and, in addition to other things, extended Dall'Aglio's Theorem (Theorem 7.1) to this setting.

A copula C is an Archimedean t-norm if it admits the representation (3.3) with f convex (see Theorem 3.2 ff.). In [24], C. Genest and R. J. MacKay initiated the study of such *Archimedean copulas*, assuming that the generator f in (3.3) is continuously twice differentiable with $f'(t) < 0$ and $f''(t) > 0$ for all t in $(0,1)$. This family includes the copulas of such standard bivariate d.f.'s as those associated with the names of Gumbel, Cook and Johnson, and Ali, Mikhail and Haq, as well as Frank's family of copulas (see (4.15)), but not Plackett's family and also not the various families of Farlie-Gumbel-Morgenstern distributions. Genest and MacKay showed that many properties of and relations among Archimedean copulas are determined by properties of their generators. For example, the Archimedean copula C has a singular component if and only if $f(0)/f'(0) \neq 0$, in which case this singular component is concentrated on the curve $f(x) + f(y) = f(0)$ and has probability mass equal to $-f(0)/f'(0)$; and Kendall's τ is given by

(7.7) $\tau(C) = 4\int_0^1 (f(t)/f'(t))dt + 1$,

and is thus linearly related to the area above the graph of f'/f.
Genest and MacKay also gave conditions on the generators f_1 and f_2
of two Archimedean copulas C_1 and C_2 which guarantee that $C_1 \leq C_2$
and showed that under these conditions $\rho(C_1) \leq \rho(C_2)$, $\tau(C_1) \leq \tau(C_2)$
and $q(C_1) \leq q(C_2)$; and for a convergent sequence of Archimedean copulas, they gave conditions under which the limit is Archimedean or is
equal to Min. In a subsequent paper [25], enticingly entitled "The Joy
of Copulas", they discussed some of the above results from a pedagogi-
cal point of view.

In [23] and in [50], Genest and R. B. Nelsen independently considered Frank's family from a statistical point of view. They showed
that it satisfies a set of criteria laid down by Kimeldorf and Sampson
in [33]: namely, all the copulas in this family are absolutely continuous and have full support; the family is linearly ordered and continuous in the parameter s; and it contains Min, Prod and W as limiting
cases. In addition, the measures of association ρ, τ and q are
monotone functions of s. They also provided (different) algorithms
for the use of these copulas in simulations; and Genest considered the
question of estimating the parameter s. And in a forthcoming paper
[26], Genest and Rivest investigate the relationship between Archimedean copulas and extreme value distributions. For further discussion
of Archimedean copulas, see Genest's contribution to this volume.

We conclude this section by remarking that copulas are employed
by W. Whitt in [76], by W. Stute in [71] and [72] and, more recently,
by A. W. Marshall and I. Olkin in [43].

8. WHEN MARGINS ARE FIXED

By now the reader should be aware of the fact that any paper bear-
ing a title that contains a phrase such as "when the margins are fixed"
is a paper involving copulas. More often than not (since copulas are
not yet as well-known as they might be) this involvement is implicit.
Consequently, there is generally something to be gained - insight at
the very least - by bringing the role of copulas explicitly to the fore.
To illustrate, in 1983, while browsing through the Mathematical Reviews,

I came across a review of a paper by G. D. Makarov entitled "Estimates
for the distribution function of the sum of two random variables with
given marginal distributions" [42; MR 83c: 60029]. I looked up the
paper and my interest in it increased when I learned that its aim was
to answer a question that had been posed by Kolmogorov. But I found
Makarov's argument somewhat impenetrable and therefore set out to practice what I have just preached. Kolmogorov's problem is the following:

Given random variables X and Y with $df(X) = F$ and $df(Y) = G$,
find distribution functions $\underline{F}$ and $\overline{F}$ such that for all z in R,

$$\underline{F}(z) = \inf P(X + Y < z)$$

and

$$\overline{F}(z) = \sup P(X + Y < z),$$

where the infimum and supremum are taken over all possible joint distribution functions H having margins F and G. To translate this into
the language of copulas we have only to recall that if C is the copula
of X and Y, then

(8.1) $df(X + Y) = \sigma_C(F,G),$

where σ_C is given by (4.10). Thus the sets

$$\{df(X + Y)\,|\,df(X) = F \quad \text{and} \quad df(Y) = G\}$$

and

$$\{\sigma_C(F,G)\,|\,C \text{ is a copula}\}$$

are identical, whence it follows that in order to determine $\underline{F}$ and $\overline{F}$
we need to find bounds on σ_C, considered as a binary operation on $\mathcal{D}$.

Having gotten this far, and being occupied by other matters (Sklar
and I were just putting the finishing touches on our book [62]), I
telephoned M. J. Frank to ask him what he knew about such bounds. He
soon returned my call and in a teasing, devilish tone of voice informed
me that the answer to my question was known. Specifically, I could
find it in a paper which should be easily accessible to me and which
had three authors: R. Moynihan, A. Sklar and ..... B. Schweizer! The

desired bounds are those given in (5.5). For in view of the fact that

$$(8.2) \qquad \tau_W(F,G) \leq \sigma_C(F,G) \leq \rho_W(F,G),$$

for all F, G in $\mathcal{D}$, with equality holding if and only if F or G
is a unit step-function, i.e., if and only if X or Y is constant
a.s., it follows that

$$(8.3) \qquad \underline{F} = \tau_W(F,G) \quad \text{and} \quad \overline{F} = \rho_W(F,G).$$

Thus Kolmogorov's problem was solved several years before it was proposed.

Among binary operations on $\mathcal{D}$, the bounds τ_W and ρ_W cannot be
improved. But more is true: For as M. J. Frank, R. B. Nelsen and I
showed in [18], these bounds cannot be improved at any point (F,G) in
$\mathcal{D} \times \mathcal{D}$, i.e., they are best-possible in the sense of the following:

THEOREM 8.1. Let F and G be any d.f.'s in $\mathcal{D}$ and let x be
any point in **R**. Then:

(i)   There exists a copula C_t, dependent only on the value t of
$\tau_W(F,G)$ at x such that

$$\sigma_{C_t} (F,G)(x) = \tau_W(F,G)(x) = t.$$

(ii)   There exists a copula C_r, dependent only on the value r of
$\rho_W(F,G)(x+)$ such that

$$\sigma_{C_r} (F,G)(x+) = \rho_W(F,G)(x+) = r .$$

For any given d.f.'s F and G and any t in **R**, there exist
r.v.'s X_t and Y_t with $df(X_t) = F$, $df(Y_t) = G$ and copula C_t,
whence $df(X_t + Y_t) = \sigma_{C_t}(F,G)$; and likewise for any r in **R** there
exist r.v.'s X_r and Y_r with $df(X_r) = F$, $df(Y_r) = G$, copula C_r
and $df(X_r + Y_r) = \sigma_{C_r}(F,G)$. This observation brings us back to Makarov who obtained the bounds in (8.2) by specifically constructing such
pairs of r.v.'s (X_t,Y_t) and (X_r,Y_r). These same results, via a
similar construction, were obtained independently and a little later
by L. Rüschendorf [53]. In establishing them, both he and Makarov made
use of the fact that for any F, G in $\mathcal{D}$ and any a in [0,1],

(8.4) $\tau_W(F,G)^{-1}(a) = \inf_{W(s,t)=a}[F^{-1}(s) + G^{-1}(t)]$

and

(8.5) $\rho_W(F,G)^{-1}(a) = \sup_{W(s,t)=a}[F^{-1}(s) + G^{-1}(t)]$,

where the inverse functions are defined by (1.9). The identities (8.4) and (8.5) are special cases of a far-reaching duality theorem establish-ed in 1979 in a joint paper with Frank [17] (see also [62, Section 7.7]).

The collection of copulas $\{C_t\}$ in (i) (resp. $\{C_r\}$ in (ii)) of Theorem 8.1 cannot be replaced by a single copula. This follows al-ready from the previously mentioned fact that equality in (8.2) holds if and only if F or G is a unit step-function. The deeper reason for this failure, however, is the fact (see Section 5) that the binary operations τ_W and ρ_W are not derivable from any binary operations on random variables.

Coming back to the paper [18], there using (8.1) and (8.3) we explicitly determined the bounds for df(X + Y) in a number of cases of special interest, namely when both F and G are uniform, exponen-tial, normal or Cauchy; and using the associativity of τ_W and ρ_W, we also showed that if $X_1,\ldots,X_n$ are r.v.'s with d.f.'s $F_1,\ldots,F_n$, re-spectively, then

(8.6) $\tau_W(F_1,\ldots,F_n) \leq df\left(\sum_{k=1}^{n} X_k\right) \leq \rho_W(F_1,\ldots,F_n)$.

Finally, again using (5.5), we obtained bounds for the d.f. of L(X,Y) for a class of functions L as well as for $L(X_1,\ldots,X_n)$ when L is associative. Subsequently, in [78] (see also [77]), R. C. Williamson and T. Downs showed that these bounds are also pointwise best-possible and used these results, together with the aforementioned duality the-orem, to develop algorithms for the effective calculation of these and other bounds on the operations $\sigma_{C,L}$.

In a later paper [51], R. B. Nelsen and I applied the results and techniques of [18] to obtain best-possible bounds for the sum of squares $X^2 + Y^2$ and the so-called radial error $(X^2 + Y^2)^{1/2}$ (which do not be-long to the class of functions L discussed above). Our central result is the following:

THEOREM 8.2. If the random variables X and Y are identically distributed and if their common distribution function G is symmetric about 0 and concave on $(0,\infty)$, then

(8.7) $\text{Max}(4G(\sqrt{x/2}) - 3,0) \leq df(X^2 + Y^2 \chi x) \leq 2G(\sqrt{x}) - 1$

and

(8.8) $\text{Max}(4G(x/\sqrt{2}) - 3,0) \leq df[(X^2 + Y^2)^{1/2}](x) \leq 2G(x) - 1.$

These bounds are best-possible and, of course, easily computed. Lastly, in the course of proving Theorem 8.2 we found that if C is a copula and C^* is defined on $[0,1]^2$ by

$$C^*(a,b) = C((1+a)/2,(1+b)/2) - C((1-a)/2,(1+b)/2)$$
$$-C((1+a)/2,(1-b)/2) + C((1-a)/2,(1-b)/2),$$

then C^* is also a copula. Note that $C^*(a,b)$ is the mass that C assigns to the rectangle whose sides have length a and b and which is centered at the point $(1/2,1/2)$.

9. RECENT DEVELOPMENTS

As has already been pointed out on several occasions, the binary operations $\tau_{T,L}$ and $\rho_{C,L}$ on D are generally not derivable from binary operations on random variables defined on a common probability space. In [3], C. Alsina and I showed that this same state of affairs also prevails in a situation that, for statisticians and probabilists, is much closer to home. Specifically, we proved that mixtures are not derivable, i.e., that if c is any fixed number between 0 and 1 and if ϕ_c is the binary operation defined on D via

(9.1) $\phi_c(F,G)(t) = cF(t) + (1 - c)G(t),$

then ϕ_c is not derivable from any binary operation on random variables. The proof is easy but the result is of some philosophical significance.

Two-dimensional copulas may be viewed as doubly stochastic measures on the unit square. As such they have attracted the attention

of various authors who, using techniques of functional analysis, considered the (still wide open) problem of finding and characterizing
the extreme points of this convex set (see, e.g., the contribution of
V. Beneš and J. Štěpán to this volume). Recently, Sherwood and Taylor [67] came to this problem from a different direction. Motivated
by the idea of redistributing the mass of Min from the main diagonal
of the unit square onto two curves joining the corners (0,0) and (1,1),
they defined a *hairpin* as the graph of $g \cup g^{-1}$, where g is an increasing homeomorphism of the unit interval satisfying $0 < g(u) < u$
for $0 < u < 1$ and g^{-1} is the inverse of g. Then, using an approach that employs functional equations, they found necessary conditions and sufficient conditions on g for the hairpin $g \cup g^{-1}$ to
contain the support of a doubly stochastic measure (copula). They
further showed that $g \cup g^{-1}$ can support at most one such measure,
whence it follows that "hairpin copulas" are extreme points. In subsequent papers with A. Kamiński and P. Mikusiński (e.g. [28]) they
considered several different classes of hairpins. These studies led
them to the "shuffles of Min" which redistribute the mass of Min in a
different fashion. These shuffles are dense in the set of all two-
dimensional copulas; nevertheless, if X and Y are r.v.'s whose copula is a shuffle of Min, then there is a functional relationship between
X and Y. These matters are discussed in detail in the contribution
of Mikusiński, Sherwood and Taylor to this volume, where further references to the relevant literature may be found. Vitale's paper [74]
contains closely related results.

The most recent and in many ways most exciting new development is
due to W. F. Darsow, Bao Nguyen and E. T. Olsen [8]. For any 2-copula
C, let $C_{,1}$ and $C_{,2}$ denote the partial derivatives $\partial C/\partial u$ and
$\partial C/\partial v$; and for any pair of 2-copulas A and B, let $A * B$ be the
function defined on $[0,1]^2$ via

(9.2) $$(A * B)(u,v) = \int_0^1 A_{,2}(u,t) B_{,1}(t,v) dt \quad .$$

Then $A * B$ is a 2-copula, i.e., $*$ is a binary operation on the set of 2-copulas. Moreover, the operation * is associative, distributes over convex combinations, possesses an identity element (namely Min) and a null element (namely Prod), and is continuous in each place (but not jointly continuous).

The salient property of the * operation, however, is the fact that if X, Y, Z are r.v.'s having the property that X and Z are conditionally independent given Y, then

(9.3) $C_{XZ} = C_{XY} * C_{YZ}$.

This leads directly to

THEOREM 9.1. Let $\{X_t | t \in T\}$ be a real stochastic process with parameter set T and, for any s, t in T, let C_{st} be the copula of X_s and X_t. Then the transition probabilities of the process satisfy the Chapman-Kolmogorov equations if and only if

(9.4) $C_{st} = C_{su} * C_{ut}$,

for all s, u, t in T satisfying s < u < t.

Theorem 7.1 is the key to a new approach to the theory of Markov processes and to a new technique for constructing them. In the conventional approach, a Markov process is specified by an initial distribution and a family of transition probabilities satisfying the Chapman-Kolmogorov equations. The new approach proceeds by giving all the one-dimensional marginal distributions and a family of 2-copulas satisfying (9.4). It is different in principle from the conventional approach, for holding the transition probabilities fixed and varying the initial distribution necessarily varies all the marginal distributions, whereas holding the copulas of the process fixed and varying the initial distribution does not affect any other marginal distribution. Darsow, Nguyen and Olsen illustrate their technique with a number of interesting examples, including a Brownian motion process with non-Gaussian marginal distributions. Also, as is well-known, the Chapman-Kolmogorov equations are a necessary but not sufficient condition for a real stochastic process to be Markov. Using copulas, they construct an example illustrating this situation; and then, using a natural extension of (9.4) to higher dimensional copulas, they give a condition which is necessary and sufficient.

In the final part of their paper, Darsow, Nguyen and Olsen define a *Markov algebra* as a pair (A,*), where A is a compact convex subset of a real Banach space and * is a binary operation on A satisfying the properties listed after (9.2) above; and A is *symmetric* if there is a mapping T: A → A satisfying T(T(U)) = U, T(λU + (1 − λ)V) = λT(U) + (1 − λ)T(V) and T(U * V) = T(V) * T(U), for all U, V in A and 0 $\leq \lambda \leq$ 1. Examples of symmetric Markov algebras are: the set of doubly stochastic n × n matrices for any positive integer n, where * is matrix multiplication and T is the transpose, and the set of all 2-copulas, where * is given by (9.2) and T is defined via (T(C))(x,y) = C(y,x). They study these algebras from an abstract point of view and give probabilistic interpretations of their results. For example, if the copula C has a left-inverse or a right-inverse (which are unique if they exist), then C is an extreme point of A; and if X and Y are continuous r.v.'s, then there exists a Borel-measurable function f such that Y = f(X) a.s. if and only if C_{XY} is left-invertible. They also study stochastic processes in this setting and point out that their approach holds promise of capturing the Markov property of such processes in a framework as simple and perspicuous as the conventional framework for analyzing Markov chains.

REFERENCES

1. Aczel, J. (1949) Sur les opérations définies pour nombres réels, *Bull. Soc. Math. France* 76, 59–64.
2. Aczel, J. (1966) *Lectures on Functional Equations and their Applications*, Academic Press, New York.
3. Alsina, C. and Schweizer, B. (1988) Mixtures are not derivable, *Found. of Physics Letters* 1, 171–174.
4. Dall'Aglio, G., (1959) Sulla compatibilità delle fuzione di ripartizione doppia, *Rend. Mat.* 18, 385–413.
5. Dall'Aglio, G. (1960) Les fonctions extrêmes de la classe de Fréchet à 3 dimensions, *Publ. Inst. Statist. Univ. Paris* 9, 175–188.
6. Dall'Aglio, G. (1961) Osservazioni sulla convergenza in distribuzione e in probabilità, *Giorn. Ist. Ital. Attuari* 24, 94–108.
7. Dall'Aglio, G. (1972) Fréchet classes and compatibility of distribution functions, *Symposia Math.* 9, 131–150.
8. Darsow, W. F., Nguyen, Bao and Olsen, E. T., Copulas and Markov processes, to appear.

9. Deheuvels, P. (1978) Caractérisation complète des lois extrêmes multivariées et de la convergence des types extrêmes, *Publ. Inst. Statist. Univ. Paris* 23, 1-36.

10. Deheuvels, P. (1980) The decomposition of infinite order and extreme multivariate distributions, in *Asymptotic Theory of Statistical Tests and Estimation (Proc. Adv. Internat. Sympos., Univ. North Carolina, Chapel Hill, N.C., 1979)*, pp. 259-286, Academic Press, New York.

11. Deheuvels, P. (1979) La fonction de dépendance empirique et ses propriétés. Un test non paramétrique d'indépendance, *Acad. Roy. Belg. Bull. Cl Sci.* (5) 65, 274-292.

12. Deheuvels, P. (1981) Multivariate tests of independence, in *Analytical Methods in Probability Theory (Oberwolfach, 1980)*, pp. 42-50, Lecture Notes in Math., 861, Springer-Verlag, Berlin.

13. Deheuvels, P. (1983) Indépendance multivariée partielle et inégalités de Fréchet, in *Studies in Probability and Related Topics, Papers in Honour of Octav Onicescu on his 90th Birthday*, ed. by M. C. Demetrescu and M. Iosifescu, Editrice Nagard, Rome, pp. 145-155.

14. Féron, R. (1956) Sur les tableaux de corrélation dont les marges sont données, cas de l'espace à trois dimensions, *Publ. Inst. Statist. Univ. Paris* 5, 3-12.

15. Frank, M. J. (1975) Associativity in a class of operations on a space of distribution functions, *Aequationes Math.* 12, 121-144.

16. Frank, M. J. (1979) On the simultaneous associativity of F(x,y) and x + y - F(x,y), *Aequationes Math.* 19, 194-226.

17. Frank, M. J. and Schweizer, B. (1979) On the duality of generalized infimal and supremal convolutions, *Rend. Mat.* 12, 1-23.

18. Frank, M. J., Nelsen, R. B. and Schweizer, B. (1987) Best-possible bounds for the distribution of a sum - a problem of Kolmogorov, *Probab. Th. Rel. Fields* 74, 199-211.

19. Fréchet, M. (1935) Généralisations du théorème des probabilités totales, *Fund. Math.* 25, 379-387.

20. Fréchet, M. (1951) Sur les tableaux de corrélation dont les marges sont données, *Ann. Univ. Lyon* 9, Sect. A, 53-77.

21. Fréchet, M. (1957) Les tableaux de corrélation et les programmes linéaires, *Revue Inst. Int. Statist.* 25, 23-40.

22. Galambos, J. (1978) *The Asymptotic Theory of Extreme Order Statistics*, John Wiley & Sons, New York.

23. Genest, C. (1987) Frank's family of bivariate distributions, *Biometrika* 74, 549-555.

24. Genest, C. and MacKay, R. J. (1986) Copules archimédiennes et familles de lois bidimensionelles dont les marges sont données, *Canadian J. Statist.* 2, 145-159.

25. Genest, C. and MacKay, R. J. (1986) The joy of copulas: bivariate distributions with uniform marginals, *Amer. Statist.* 40, 280-283.

26. Genest, C. and Rivest, L. P. (1989) A characterization of Gumbel's family of extreme value distributions, *Statist. and Probab. Letters* 8, 207-211.

27. Hoeffding, W. (1940) Masstabinvariante Korrelationstheorie, *Schriften des Mathematischen Instituts und des Instituts für Angewandte Mathematik der Universität Berlin* 5, 179-233.
28. Kamiński, A., Mikusiński, P., Sherwood, H. and Taylor, M. D. (1988) Properties of a special class of doubly stochastic measures, *Aequationes Math.* 36, 212-229.
29. Kellerer, H. G. (1964) Masstheoretische Marginalprobleme, *Math. Ann.* 153, 168-198.
30. Kellerer, H. G.(1964) Verteilungsfunktionen mit gegebenen Marginalverteilungen, *Z. Warsch. Verw. Geb.* 3, 247-270.
31. Kimberling, C. H. (1973) Exchangeable events and completely monotonic sequences, *Rocky Mountain J. Math.* 3, 565-574.
32. Kimberling, C. H. (1974) A probabilistic interpretation of complete monotonicity, *Aequationes Math.* 10, 152-164.
33. Kimeldorf, G. and Sampson, A. R. (1975) One-parameter families of bivariate distributions with fixed marginals, *Commun. Statist.* 4, 293-301.
34. Kimeldorf, G. and Sampson, A. R. (1975) Uniform representations of bivariate distributions, *Commun. Statist.* 4, 617-627.
35. Kimeldorf, G. and Sampson, A. R. (1978) Monotone dependence, *Ann. Statist.* 6, 895-903.
36. Kimeldorf, G. and Sampson, A. R. (1987) Positive dependence orderings, *Ann. Inst. Statist. Math.* 39, 113-128.
37. Kimeldorf, G. and Sampson, A. R. (1989) A framework for positive dependence, *Ann. Inst. Statist. Math.* 41, 31-45.
38. Kotz, S. and Johnson, N. L. (1977) Propriétés de dépendance des distributions intérées généralisees à deux variables Farlie-Gumbel-Morgenstern, *C. R. Acad. Sci. Paris* 285A, 277-280.
39. Kruskal, W. H. (1958) Ordinal measures of association, *J. Amer. Statist. Assoc.* 53, 814-861.
40. Lehmann, E. L. (1966) Some concepts of dependence, *Ann. Math. Statist.* 37, 1137-1153.
41. Ling, C. H. (1965) Representation of associative functions, *Publ. Math. Debrecen* 12, 189-212.
42. Makarov, G. D. (1981) Estimates for the distribution function of a sum of two random variables when the marginal distributions are fixed, *Theor. Probab. Appl.* 26, 803-806.
43. Marshall, A. W. and Olkin, I. (1988) Families of multivariate distributions, *J. Amer. Statist. Assoc.* 83, 834-841.
44. Menger, K. (1942) Statistical metrics, *Proc. Nat. Acad. Sci. U.S.A.* 28, 535-537.
45. Moore, D. S. and Spruill, M. C. (1975) Unified large-sample theory of general chi-squared statistics for tests of fit, *Ann. of Math.* 65, 117-143.
46. Mostert, P. S. and Shields, A. L. (1957) On the structure of semigroups on a compact manifold with boundary, *Ann. Statist.* 3, 599-616.
47. Moynihan, R. (1978) On τ_T-semigroups of probability distribution functions, II, *Aequationes Math.* 17, 19-40.

48. Moynihan, R. and Schweizer, B. (1979) Betweenness relations in probabilistic metric spaces, *Pacific J. Math.* 81, 175-196.
49. Moynihan, R., Schweizer, B. and Sklar, A. (1978) Inequalities among binary operations on probability distribution functions, in *General Inequalities 1*, ed. by E. F. Beckenbach, Birkhäuser Verlag,Basel, pp. 133-149.
50. Nelsen, R. B., (1986) Properties of a one-parameter family of bivariate distributions with specified marginals, *Commun. Statist. - Theory Meth.* 15, 3277-3285.
51. Nelsen, R. B. and Schweizer, B., Bounds on distribution functions for sums of squares and radial errors, to appear.
52. Rényi, A. (1959) On measures of dependence, *Acta Math. Acad. Sci. Hungar.* 10, 441-451.
53. Rüschendorf, L. (1982) Random variables with maximum sums, *Adv. Appl. Probab.* 14, 623-632.
54. Rüschendorf, L. (1985) Construction of multivariate distributions with given marginals, *Ann Inst. Statist. Math.* 37, 225-233.
55. Scarsini, M. (1984) On measures of concordance, *Stochastica* 8, 201-218.
56. Scarsini, M. (1984) Strong measures of concordance and convergence in probability, *Riv. Mat. Sci. Econom. Soc.* 7, 39-44.
57. Scarsini, M. (1988) Multivariate stochastic dominance with fixed dependence structure, *Operations Res. Letters* 7, 237-240.
58. Scarsini, M., Copulae of probability measures on product spaces, to appear.
59. Schweizer, B. and Sklar, A. (1958) Espaces métriques aléatoires, *C. R. Acad. Sci. Paris* 247, 2092-2094.
60. Schweizer, B. and Sklar, A. (1961) Associative functions and statistical triangle inequalities, *Publ. Math. Debrecen* 8, 169-186.
61. Schweizer, B. and Sklar, A. (1974) Operations on distribution functions not derivable from operations on random variables, *Studia Math.* 52, 43-52.
62. Schweizer, B. and Sklar, A. (1983) *Probabilistic Metric Spaces*, Elsevier North-Holland, New York.
63. Schweizer, B. and Wolff, E. F. (1976) Sur une mesure de dépendance pour les variables aléatoires, *C. R. Acad. Sci. Paris* 283A, 659-661.
64. Schweizer, B. and Wolff, E. F. (1981) On nonparametric measures of dependence for random variables, *Ann Statist.* 9, 879-885.
65. Šerstnev, A. N. (1963) On the notion of a random normed space, *Dokl. Akad. Nauk SSSR* 149, 280-283.
66. Šerstnev, A. N. (1964) On a probabilistic generalization of metric spaces, *Kazan Goz. Univ. Učen. Zap.* 124, 3-11.
67. Sherwood, H. and Taylor, M. D. (1988) Doubly stochastic measures with hairpin support, *Probab. Th. Rel. Fields* 78, 617-626.
68. Sklar, A. (1959) Fonctions de répartition à n dimensions et leurs marges, *Publ. Inst. Statist. Univ. Paris* 8, 229-231.
69. Sklar, A. (1973) Random variables,joint distribution functions and copulas, *Kybernetika* 9, 449-460.

70. Strassen, V. (1965) The existence of probability measures with given marginals, *Ann. Math. Statist.* 36, 423-439.

71. Stute, W. (1986) Conditional empirical processes, *Ann. Statist.* 14, 638-647.

72. Stute, W. (1986) On almost sure convergence of conditional empirical distribution functions, *Ann. Statist.* 14, 891-901.

73. Urbanik, K. (1964, 1973, 1984, 1986) Generalized convolutions I, II, III and IV, *Studia Math.* 23, 217-245; 45, 57-70; 80, 167-189; and 83, 57-95.

74. Vitale, R., Stochastic dependence and a class of degenerate distributions, in *Topics in Statistical Dependence*, ed. by H. Block, A. R. Sampson and T. Savits, IMS Lecture Notes and Monograph Series, to appear.

75. Wald, A. (1943) On a statistical generalization of metric spaces, *Proc. Nat. Acad. Sci. U.S.A.* 29, 196-197.

76. Whitt, W. (1976) Bivariate distributions with given marginals, *Ann. Statist.* 4, 1280-1289.

77. Williamson, R. C., An extreme limit theorem for dependency bounds of normalized sums of random variables, *Information Sciences*, to appear.

78. Williamson, R. C. and Downs, T. (1990) Probabilistic arithmetic: numerical methods for calculating convolutions and dependency bounds, *Int. J. Approximate Reasoning* 4, 89-158.

79. Wolff, E. F. (1977) Measures of dependence derived from copulas, Ph.D. Thesis, Univ. Massachusetts, Amherst.

80. Wolff, E. F. (1981) N-dimensional measures of dependence, *Stochastica* 4, 175-188.

COPULAS AND ASSOCIATION

ROGER B. NELSEN
Department of Mathematics
Lewis and Clark College
Portland, Oregon 97219 U.S.A.

ABSTRACT. A copula is a function of two variables which couples a bivariate distribution function to its marginal distribution functions. In doing so the copula captures certain nonparametric aspects of the relationship between the variates, from which it follows that measures of association and positive dependence concepts are properties of the copula. In this paper we survey results relating copulas to Spearman's rho and Kendall's tau for a variety of bivariate distributions, and we also show that certain positive dependence concepts (positive quadrant dependent, right tail increasing, left tail decreasing, and stochastically increasing) can be interpreted as simple geometric properties of the copula.

1. Introduction

If two random variables X and Y are not independent then they are said to be "dependent," "correlated," or "associated." Much of the statistical literature is devoted to describing or measuring this association. In this paper we will survey some results concerning two measures of association and several positive dependence properties and their relationship to the copula for the joint distribution of X and Y.

We will assume that the reader is familiar with the basic properties of copulas as discussed by Schweizer and Sklar (1983) and in Schweizer's contribution to these proceedings. Throughout this paper we will assume that the random variables X and Y are continuous with joint distribution function H and marginal distribution functions F and G. We will let C (subscripted if necessary) denote the copula of X and Y, so that $H(x,y)=C(F(x),G(y))$. If we let U and V denote the uniform random variables F(X) and G(Y) respectively, then the joint distribution function for U and V is C. The copulas of the upper and lower Fréchet bounds will be denoted by M and W where $M(u,v) = \min(u,v)$

and $W(u,v) = \max(0,u+v-1)$ respectively, and Π will denote the copula of independent random variables so that $\Pi(u,v) = uv$. Survival functions will be denoted $\overline{F}$, $\overline{G}$, and $\overline{H}$; and I^2 will denote the unit square $[0,1]\times[0,1]$. When required, we will make additional assumptions concerning the existence of the first-order partial derivatives of C on $(0,1)^2$.

2. Measures of Association

Originally created to measure association in samples, the nonparametric correlation coefficients commonly known as Kendall's tau (τ) and Spearman's rho (ρ) have population analogs which can be expressed in terms of the copula C. Since the term "correlation" usually refers to a measure of the linear relationship between variables, τ and ρ are today often referred to as measures of association.

2.1. KENDALL'S τ

This measure of association was first discussed by G. T. Fechner around 1900, and rediscovered by M. G. Kendall in 1938. For a complete historical review of τ (and ρ), see Kruskal (1958).

We say that two pairs of random variables (X_1,Y_1) and (X_2,Y_2) are *concordant* if $X_1<X_2$ and $Y_1<Y_2$ or if $X_1>X_2$ and $Y_1>Y_2$; and *discordant* if $X_1<X_2$ and $Y_1>Y_2$ or if $X_1>X_2$ and $Y_1<Y_2$. Then Kendall's τ can be defined as the difference between the probabilities of concordance and discordance for two independent pairs (X_1,Y_1) and (X_2,Y_2) each with distribution H; that is

$$\tau = \Pr\{(X_1 - X_2)(Y_1 - Y_2) > 0\} - \Pr\{(X_1 - X_2)(Y_1 - Y_2) < 0\}. \tag{2.1}$$

But these probabilities can be evaluated by integrating over the distribution of (X_2,Y_2), so that writing (X,Y) for (X_1,Y_1), we have

$$\tau = \iint_{R^2} [\Pr\{X<x,Y<y\} + \Pr\{X>x,Y>y\} - \Pr\{X<x,Y>y\} - \Pr\{X>x,Y<y\}]dH,$$

$$= \iint_{R^2} \{H(x,y) + [1-F(x)-G(y)+H(x,y)] - [F(x)-H(x,y)] - [G(y)-H(x,y)]\}dH,$$

$$= \iint_{I^2} \{C(u,v) + [1 - u - v + C(u,v)] - [u - C(u,v)] - [v - C(u,v)]\}dC,$$

where the third integral is obtained from the second via the substitution $u = F(x)$, $v = G(y)$. Hence we have

$$\tau = 4 \iint_{I^2} C(u,v)dC - 1. \qquad (2.2)$$

Other derivations of this representation for τ are also possible [see Conway (1979) and Kruskal (1958)].

However, the representation for τ given by (2.2) is often not amenable to computation, especially for copulas with both an absolutely continuous and a singular component. For many such calculations, the expression

$$\tau = 1 - 4 \iint_{I^2} \frac{\partial}{\partial u}C(u,v)\frac{\partial}{\partial v}C(u,v) \, dudv \qquad (2.3)$$

is more tractable. To establish the equivalence of (2.2) and (2.3), we need only prove

Theorem 2.1 If C is a copula, then

$$\iint_{I^2} C(u,v) \, dC + \iint_{I^2} \frac{\partial}{\partial u}C(u,v)\frac{\partial}{\partial v}C(u,v) \, dudv = \frac{1}{2}. \qquad (2.4)$$

Proof. We will assume that for every $v \in (0,1)$, $\frac{\partial}{\partial u}C(u,v)$ fails to exist for at most finitely many values of $u \in (0,1)$; and similarly for $\frac{\partial}{\partial v}C(u,v)$. [This will be sufficient for the families of copulas to be considered in Section 3.] Let P denote a partition of I^2 into subrectangles $R_{i,j}$ formed in the following manner: (i) partition [0,1] on the v-axis to induce a partition of I^2 into horizontal strips $[0,1] \times [v_{j-1},v_j]$, $1 \leq j \leq m$ with $v_0 = 0$, $v_m = 1$; (ii) choose points $\{u_{i,j}: 1 \leq i \leq n_j\}$ [with $u_{0,j} = 0$, $u_{n_j,j} = 1$] to partition the jth strip into subrectangles so that the only points on the grid determined by P (i.e., the boundaries of the $R_{i,j}$'s) where either partial fails to exist are the lattice points of that grid. Let $\|P\|$ denote the norm of P and let $\Delta u_{i,j}$ and Δv_j denote the dimensions of $R_{i,j}$.

Applying the mean value theorem to $C(u,v)$ for $u \in [u_{i-1,j}, u_{i,j}]$, $v = v_j$, and to $C(u,v)$ for $v \in [v_{j-1},v_j]$, $u = u_{i-1,j}$; we obtain $C(u_{i,j},v_j) - C(u_{i-1,j},v_j) = \frac{\partial}{\partial u}C(u^*_{i,j},v_j)\Delta u_{i,j}$ where $u^*_{i,j} \in$

$(u_{i-1,j}, u_{i,j})$, and $C(u_{i-1,j}, v_j) - C(u_{i-1,j}, v_{j-1}) = \frac{\partial}{\partial v} C(u_{i-1,j}, v_{i,j}^*) \Delta v_j$ where $v_{i,j}^* \in (v_{j-1}, v_j)$. But

by Duhamel's principle,

$$\iint_{I^2} \frac{\partial}{\partial u} C(u,v) \frac{\partial}{\partial v} C(u,v) \, du dv$$

$$= \lim_{\|P\| \to 0} \sum_{j=1}^{m} \sum_{i=1}^{n_j} \frac{\partial}{\partial u} C(u_{i,j}^*, v_j) \frac{\partial}{\partial v} C(u_{i-1,j}, v_{i,j}^*) \Delta u_{i,j} \Delta v_j,$$

$$= \lim_{\|P\| \to 0} \sum_{j=1}^{m} \sum_{i=1}^{n_j} [C(u_{i,j}, v_j) - C(u_{i-1,j}, v_j)][C(u_{i-1,j}, v_j) - C(u_{i-1,j}, v_{j-1})].$$

We also have $\iint_{I^2} C dC$

$$= \lim_{\|P\| \to 0} \sum_{j=1}^{m} \sum_{i=1}^{n_j} C(u_{i,j}, v_j)[C(u_{i,j}, v_j) - C(u_{i-1,j}, v_j) - C(u_{i,j}, v_{j-1}) + C(u_{i-1,j}, v_{j-1})],$$

and thus the left side of (2.4) can be written (following some simplification) as

$$\lim_{\|P\| \to 0} \sum_{j=1}^{m} \sum_{i=1}^{n_j} [C^2(u_{i,j}, v_j) - C(u_{i,j}, v_j) C(u_{i,j}, v_{j-1}) - C^2(u_{i-1,j}, v_j) + C(u_{i-1,j}, v_j) C(u_{i-1,j}, v_{j-1})].$$

But the inner sum telescopes to $C^2(u_{nj,j}, v_j) - C(u_{nj,j}, v_j) C(u_{nj,j}, v_{j-1}) = v_j^2 - v_j v_{j-1} = v_j \Delta v_j$

and thus the left side of (2.4) reduces to

$$\lim_{\|P\| \to 0} \sum_{j=1}^{m} v_j \Delta v_j = \int_0^1 v \, dv = \frac{1}{2},$$

completing the proof.

Since $W(u,v) \le C(u,v) \le M(u,v)$, it follows that $\iint_{I^2} W dW \le \iint_{I^2} C dW$ and $\iint_{I^2} C dM \le \iint_{I^2} M dM$. But since C is continuous and 2-increasing, we also have $\iint_{I^2} C dW \le \iint_{I^2} C dC \le \iint_{I^2} C dM$ [Whitt (1976); Conway (1979); Tchen (1980)], and hence $\tau_W \le \tau \le \tau_M$ [where τ_W and τ_M denote Kendall's τ for W and M, respectively]. Furthermore, from (2.3), we have

$$\tau_W = 1 - \int_0^1 \int_{1-u}^1 1 \, dv du = -1 \quad \text{and} \quad \tau_M = 1 - \int_0^1 \int_0^1 0 \, du dv = +1.$$

Thus $-1 \leq \tau \leq 1$, and so τ represents a "scaled" expected value of C.

2.2. SPEARMAN'S ρ

First proposed by the psychologist C. Spearman in 1904, this coefficient is also known as the *grade correlation coefficient,* a term introduced by K. Pearson. Like Kendall's τ, it is related to the difference between probabilities of concordance and discordance. The distinction is that here one pair, say (X,Y), has distribution H, while the second pair, say (X',Y') is a pair of *independent* r.v.'s with the same margins as X and Y; i.e., (X',Y') has distribution function $F(x)G(y)$. Equivalently, we can consider *three* independent pairs (X_1,Y_1), (X_2,Y_2), and (X_3,Y_3) each with distribution function H, and consider the probability p that (X_1,Y_1) and (X_2,Y_3) are concordant.

Let $p = \Pr\{(X_1 - X_2)(Y_1 - Y_3) > 0\}$, that is

$$p = \Pr\{X_1 > X_2, Y_1 > Y_3\} + \Pr\{X_1 < X_2, Y_1 < Y_3\}. \tag{2.5}$$

If we integrate with respect to the distribution of (X_1,Y_1) and recall that X_2 and Y_3 are independent, we have

$$\begin{aligned}
p &= \iint_{R^2} [\Pr\{X < x\}\Pr\{Y < y\} + \Pr\{X > x\}\Pr\{Y > y\}] \, dH, \\
&= \iint_{R^2} [F(x)G(y) + (1 - F(x))(1 - G(y))] \, dH, \\
&= \iint_{I^2} [uv + (1 - u)(1 - v)] \, dC, \\
&= 2 \iint_{I^2} uv \, dC \; = \; 2E(UV) = 2E(F(X)G(Y)).
\end{aligned}$$

Hence p (and consequently ρ) is a function of Pearson's product-moment correlation coefficient r for the random variables U and V. If we define Spearman's ρ for (X,Y) to be Pearson's r for $(U,V) = (F(X),G(Y))$, then we will have $-1 \leq \rho \leq 1$ (as with τ). Thus

$$\rho \; = \; \frac{E(UV) - \frac{1}{4}}{\frac{1}{12}} \; = \; 12E(UV) - 3 \; = \; 12 \iint_{I^2} uv \, dC - 3. \tag{2.6}$$

An alternate form for ρ can be obtained from (2.5) by integrating with respect to the distribution of (X_2,Y_3). Thus

$$p = \iint_{R^2} [Pr\{X>x,Y>y\} + Pr\{X<x,Y<y\}] \, dFdG,$$

$$= \iint_{R^2} \{[1 - F(x) - G(y) + H(x,y)] + H(x,y)\} \, dFdG,$$

$$= \iint_{I^2} \{[1 - u - v + C(u,v)] + C(u,v)\} \, dudv,$$

$$= 2 \iint_{I^2} C(u,v) \, dudv.$$

Scaling as before, we have

$$\rho = 12 \iint_{I^2} C(u,v) \, dudv - 3. \tag{2.7}$$

From (2.7) we obtain several geometric interpretations of the coefficient ρ. It can be viewed either as the volume under the surface $z=C(u,v)$ over I^2, scaled to lie in the interval $[-1,1]$; or as the mean height of that surface so scaled. Alternatively, we have $\rho=12\iint_{I^2}[C(u,v)-uv] \, dudv$, so that ρ represents the (scaled) signed volume between the surfaces $z = C(u,v)$ and $z = \Pi(u,v) = uv$ [see Schweizer and Wolff, 1981].

2.3. THE RELATIONSHIP BETWEEN τ AND ρ

The relationship between τ and ρ, and between their sample estimates, has long been studied [see Kruskal (1958) and Kendall (1962)]. The expressions given by (2.2) for τ and (2.6) for ρ might suggest that there is a functional relationship between the two. Such is not the case; but sharp inequalities relating τ and ρ do exist.

   Daniels (1950) showed that $-1 \leq 3\tau - 2\rho \leq 1$, and Durbin and Stuart (1951) subsequently refined this to $\frac{3}{2}\tau - \frac{1}{2} \leq \rho \leq \frac{1}{2} + \tau - \frac{1}{2}\tau^2$ for $\tau \geq 0$ and $\frac{1}{2}\tau^2 + \tau - \frac{1}{2} \leq \rho \leq \frac{3}{2}\tau + \frac{1}{2}$ for $\tau \leq 0$. Thus for any copula C, the values of τ and ρ must lie in the shaded region in Figure 1. We shall refer to this region as the τ,ρ-region.

3. Examples

We now study the relationship between τ and ρ for selected families of copulas. For each family, τ and ρ have been evaluated using the expressions given in the preceding section.

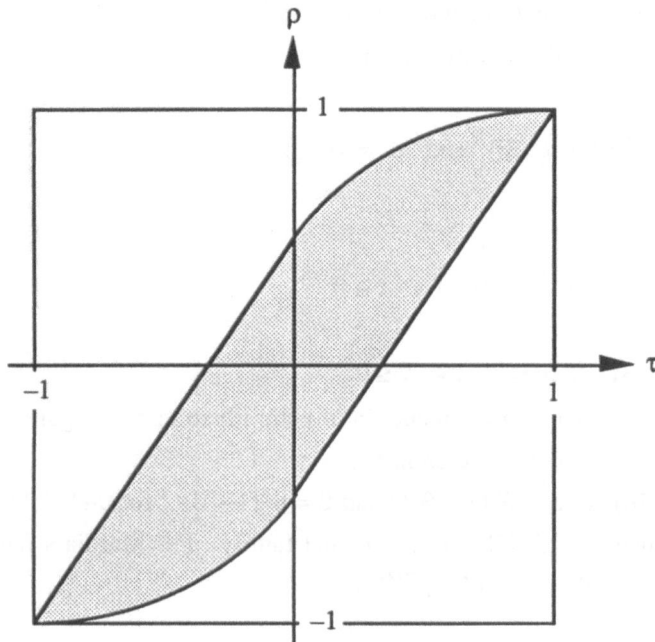

Figure 1. The τ,ρ-region.

While these calculations are not difficult, they are tedious, and have been suppressed. Many were done using the symbolic algebra program MAPLE® on a Macintosh Plus computer. While the difference between τ and ρ can be as much as 0.5 for some copulas, we shall see that for many of these families, there is nearly a functional relationship between the two.

For surveys of additional families of copulas and the associated values of τ and ρ, see Conway (1979) and Hutchinson and Lai (1990).

3.1. FRÉCHET'S FAMILY

Fréchet (1958) proposed a two-parameter convex linear combination of Π and the bounds M and W, namely

$$C_{\alpha,\beta}(u,v) = \alpha M(u,v) + (1-\alpha-\beta)uv + \beta W(u,v); \quad \alpha,\beta \geq 0, \quad \alpha + \beta \leq 1. \qquad (3.1)$$

[A family of copulas which includes M, W, and Π is called *inclusive* or *comprehensive* (Devroye, 1986).] For this family, we have

$$\tau = \frac{\alpha - \beta}{3}(2 + \alpha + \beta) \quad \text{and} \quad \rho = \alpha - \beta.$$

It follows that
$$\tau \le \rho \le -1 + \sqrt{1 + 3\tau} \quad \text{for} \quad \tau \ge 0,$$
and
$$1 - \sqrt{1 + 3\tau} \le \rho \le \tau \quad \text{for} \quad \tau \le 0.$$

These bounds for (τ, ρ) for the Fréchet family are illustrated in Figure 2 (along with the boundary of the τ, ρ-region for reference).

Mardia (1970) set $\alpha = \theta^2(1 + \theta)/2$ and $\beta = \theta^2(1 - \theta)/2$ for $\theta \in [-1,1]$ to obtain a one-parameter inclusive subfamily of the Fréchet family. For Mardia's family, $\rho = \theta^3$ and $\tau = \theta^3(2 + \theta^2)/3$, so that $\tau = \rho(2 + \rho^{2/3})/3$.

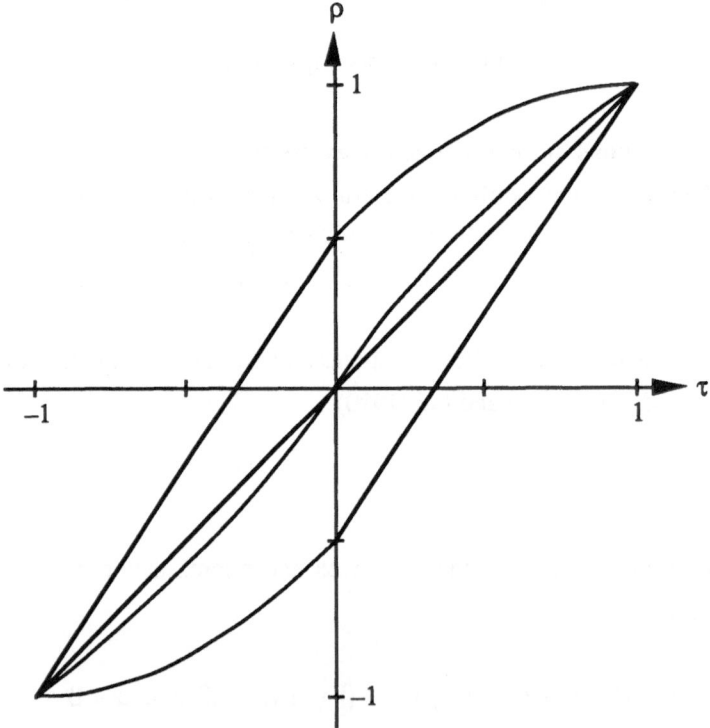

Figure 2. Bounds for ρ and τ for Fréchet's family.

3.2. GENERALIZED CUADRAS-AUGÉ COPULAS

If we set $\beta=0$ in (3.1), we obtain a subfamily of copulas with non-negative correlation given by $\alpha\min(u,v) + (1 - \alpha)uv$, $\alpha\in [0,1]$. Note that each is a weighted arithmetic mean of M and Π. If we consider weighted *geometric* means of M and Π, we have the Cuadras-Augé (1981) family ($\theta\in [0,1]$):

$$C_\theta(u,v) = [\min(u,v)]^\theta(uv)^{1-\theta} = \begin{cases} uv^{1-\theta}, & u\leq v, \\ u^{1-\theta}v, & u\geq v. \end{cases} \qquad (3.2)$$

Before discussing ρ and τ for this family, we first generalize it to a two parameter family which contains (3.2) as a subfamily. For $\alpha,\beta\in [0,1]$, define

$$C_{\alpha,\beta}(u,v) = u^{1-\alpha}v^{1-\beta}\min(u^\alpha,v^\beta) = \min(u^{1-\alpha}v,uv^{1-\beta}). \qquad (3.3)$$

Note that $C_{\alpha,0} = C_{0,\beta} = \Pi$, $C_{1,1} = M$, and for $\alpha = \beta = \theta$, (3.3) reduces to (3.2). For the family given by (3.3), we have

$$\tau = \frac{\alpha\beta}{\alpha - \alpha\beta + \beta} \quad \text{and} \quad \rho = \frac{3\alpha\beta}{2\alpha - \alpha\beta + 2\beta}$$

so that

$$\rho = \frac{3\tau}{2 + \tau}.$$

$C_{\alpha,\beta}$ is not absolutely continuous when $\alpha,\beta\in (0,1]$, as it assigns positive probability mass to the curve $v^\beta = u^\alpha$. The probability mass assigned to that curve is

$$\Pr\{U^\alpha = V^\beta\} = 1 - \iint_{I^2} \frac{\partial^2}{\partial u\partial v}C_{\alpha,\beta}(u,v)\,dudv = \frac{\alpha\beta}{\alpha - \alpha\beta + \beta} = \tau.$$

Although each copula $C_{\alpha,\beta}$ has a singular component for $\alpha,\beta\in (0,1]$, there are applications of these copulas. Each $C_{\alpha,\beta}$ is the *survival copula* (Muliere and Scarsini, 1987) for a member of the Marshall-Olkin (1967) bivariate exponential family. [The survival copula for a joint d.f. H is a copula $C_{\overline{H}}$ which satisfies $\overline{H}(x,y)=C_{\overline{H}}(\overline{F}(x),\overline{G}(y))$.] In the Marshall-Olkin family, $X \sim \exp(\lambda_1 + \lambda_{12})$, $Y \sim \exp(\lambda_2 + \lambda_{12})$, so that $\overline{F}(x) = \exp(-(\lambda_1+\lambda_{12})x)$ and $\overline{G}(y) = \exp(-(\lambda_2+\lambda_{12})y)$. Now let $\alpha = \lambda_{12}/(\lambda_1+\lambda_{12})$ and $\beta = \lambda_{12}/(\lambda_2+\lambda_{12})$, and use $C_{\alpha,\beta}$ from (3.3) for $C_{\overline{H}}$. Then

$$\overline{H}(x,y) = \exp(-\lambda_1 x)\exp(-\lambda_2 y)\min[\exp(-\lambda_{12}x),\exp(-\lambda_{12}y)]$$
$$= \exp(-\lambda_1 x - \lambda_2 y - \lambda_{12}\max(x,y)),$$

as required.

A copula C' defined by $C'(u,v) = u + v - 1 + C(1 - u, 1 - v)$ is called the *complement* of C. Since $C'' = C$, it follows that the copula C_H [i.e., $H(x,y) = C_H(F(x),G(y))$] is the complement of the survival copula $C_{\overline{H}}$, and hence the copula for each member of the Marshall-Olkin family is given by $u + v - 1 + C_{\alpha,\beta}((1 - u, 1 - v)$ with the above values for α and β. [Also see Conway (1979).]

3.3. RAFTERY'S BIVARIATE FAMILY

Raftery (1984, 1985) introduced a one-parameter $(\theta \in [0,1])$ family of absolutely continuous (for $\theta \neq 1$) bivariate distributions with identically distributed exponential margins. The survival copulas for the Raftery family are given by

$$C_\theta(u,v) = \begin{cases} u - \dfrac{1-\theta}{1+\theta}u^{\frac{1}{1-\theta}}\left[v^{\frac{-\theta}{1-\theta}} - v^{\frac{1}{1-\theta}}\right], & u \leq v; \\[3mm] v - \dfrac{1-\theta}{1+\theta}v^{\frac{1}{1-\theta}}\left[u^{\frac{-\theta}{1-\theta}} - u^{\frac{1}{1-\theta}}\right], & u > v. \end{cases}$$

$C_0 = \Pi$, $C_1 = M$. For this family we have

$$\tau = \frac{2\theta}{3-\theta} \quad \text{and} \quad \rho = \frac{\theta(4-3\theta)}{(2-\theta)^2},$$

so that

$$\rho = \frac{3\tau(8-5\tau)}{(4-\tau)^2}.$$

The curves relating τ and ρ for the Cuadras-Augé and Raftery families are illustrated in Figure 3. The Raftery family can be extended to an inclusive family by defining $C_\theta(u,v)$ for $\theta \in [-1,0)$ to be either $v - C_{-\theta}(1-u,v)$ or $u - C_{-\theta}(u,1-v)$.

3.4. A NEW FAMILY OF COPULAS

Closely related to the family given by (3.3) is the family

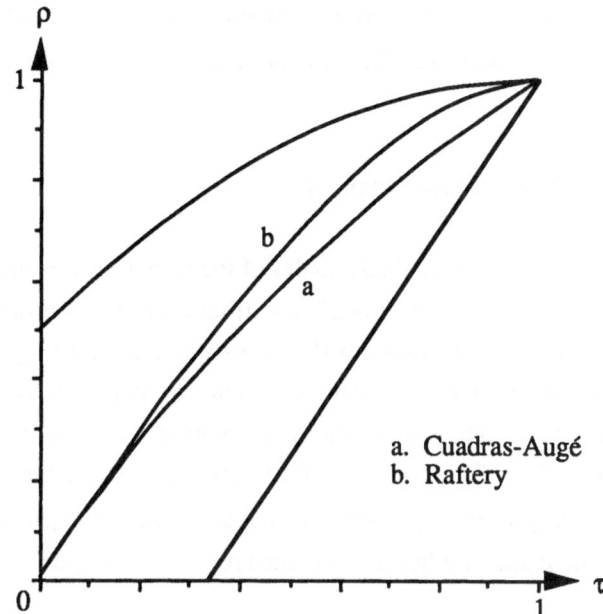

Figure 3. τ and ρ for the Cuadras-Augé and Raftery families.

$$C_{\alpha,\beta}(u,v) = \min[u, v, u^{1-\alpha}v^{1-\beta}]; \quad \alpha,\beta \geq 0, \ \alpha + \beta \leq 1. \tag{3.4}$$

Note that $C_{0,0} = \Pi$, $C_{\alpha,1-\alpha} = M$, and when $\alpha = 0$ or $\beta = 0$, the copula is also a member of the generalized Cuadras-Augé family [$\alpha=1$ or $\beta=1$, respectively, in (3.3)]. For $\alpha,\beta>0$, $\alpha+\beta<1$, the copulas given by (3.4) distribute the probability mass on and between the curves $v^{\beta} = u^{1-\alpha}$ and $v^{1-\beta} = u^{\alpha}$. For this family we have

$$\tau = \alpha + \beta \quad \text{and} \quad \rho = \frac{3(2\alpha + 2\beta - 2\alpha^2 - 2\beta^2 - \alpha\beta)}{(2 - 2\alpha + \beta)(2 - 2\beta + \alpha)}.$$

Thus

$$\frac{3\tau}{2+\tau} \leq \rho \leq \frac{3\tau(8 - 5\tau)}{(4 - \tau)^2}.$$

Hence these copulas have values for (τ,ρ) that lie on and between the curves for the Cuadras-Augé and Raftery families as shown in Figure 3.

As with the generalized Cuadras-Augé copulas, the copulas given by (3.4) have both a singular and an absolutely continuous component for $\alpha,\beta > 0$, $\alpha + \beta < 1$. The probability mass on the curve $v^{1-\beta} = u^\alpha$ is α, the probability mass on the curve $v^\beta = u^{1-\alpha}$ is β, so once again the probability mass on the singular component is numerically equal to Kendall's τ.

3.5. SHERWOOD-TAYLOR SINGULAR COPULAS

Sherwood and Taylor (1988) have studied copulas of the form $C(u,v) = \min[u,v,f(u)+f(v)]$ (also see Mikusiński, Sherwood, and Taylor's contribution to these proceedings). Such copulas assign all the probability mass to the curves $v=g(u)$ and $v=g^{-1}(u)$ where the functions $f:I\to[0,1/2]$ and $g:I\to I$ are increasing homeomorphisms satisfying the functional equation $g(x)=f(x)+f(g(x))$. They show that the function f (which they call a *mass-spreader*) satisfies $\max[0,u-1/2] < f(u) < u/2$ for $u\in (0,1)$ and g satisfies $\max[0,u-1/2] < g(u) < u$ (the set $\{(u,v)|v=g(u)$ or $v=g^{-1}(u)\}$ is called a *hairpin*). The graph of f lies in the shaded region A and the graph of g lies in the shaded region B of Figure 4.

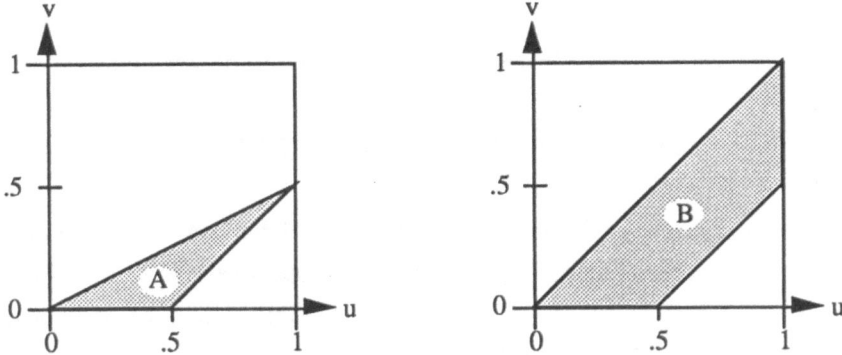

Figure 4. Regions for mass-spreaders and hairpins for Sherwood-Taylor copulas.

For such copulas, it is easily shown that

$$\tau = 8\int_0^1 f(u)\,du - 1 \quad \text{and} \quad \rho = 24\int_0^1 u\,g(u)\,f'(u)\,du - 3.$$

We can create a simple two-parameter family of these copulas by chosing any point (a,b) in region A and letting f be the function whose graph is the polygonal line joining $(0,0)$ to (a,b) to $(0,\frac{1}{2})$. Then g is the function whose graph is the polygonal line joining $(0,0)$ to $\left(a, \dfrac{ab}{a-b}\right)$ to $\left(1 - \dfrac{(1-a)(1-2a+2b)}{1-2b}, a\right)$ to $(1,1)$; similarly for g^{-1}. For example, when $(a,b) = (\frac{1}{2}, \frac{1}{6})$ the copula min[u, v, f(u) + f(v)] concentrates the probability mass on the graphs of g and g^{-1} shown in Figure 5.

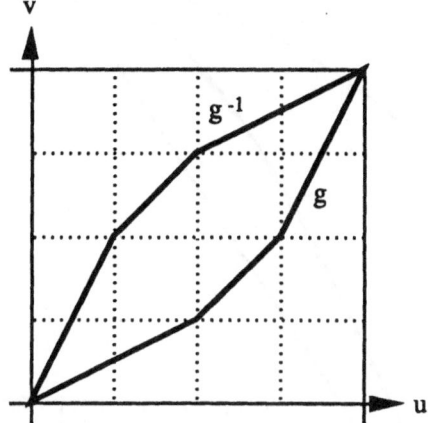

Figure 5. Graphs of g and g^{-1} when $(a,b) = (\frac{1}{2}, \frac{1}{6})$.

The expressions for τ and ρ are somewhat simpler if we reparametrize by setting $a = (1 - \alpha + \beta)/2$ and $b = \beta/2$ where $\alpha, \beta \geq 0$ and $\alpha + \beta \leq 1$. Then for this family we have

$$\tau = \alpha + \beta \quad \text{and} \quad \rho = 1 - 2(1 - \alpha - \beta)^2 + \frac{(1 - \alpha - \beta)^4}{2(1 - \alpha)(1 - \beta)},$$

from which it follows that

$$4\tau + 3 + \frac{8\tau - 14}{(2 - \tau)^2} \leq \rho \leq \frac{1}{2}(-1 + 5\tau - \tau^2 - \tau^3).$$

These bounds on (τ,ρ) for this family are remarkably narrow and almost coincide, as shown in Figure 6. The maximum vertical separation is $(533\sqrt{13} - 1921)/108 \cong 0.00703$, and occurs at $\tau = (5 - \sqrt{13})/3$. [A curious member of this family is the copula corresponding to $\alpha = \beta = (3 - \sqrt{3})/6$, for which the graphs of g and g^{-1} are arcs of regular dodecagons, and $\rho = \tau$.]

3.6. SHUFFLES OF MIN

An examination of Figures 2,3, and 6 shows that for the families of copulas considered above, the values of ρ and τ are concentrated well within the interior of the τ,ρ-region.

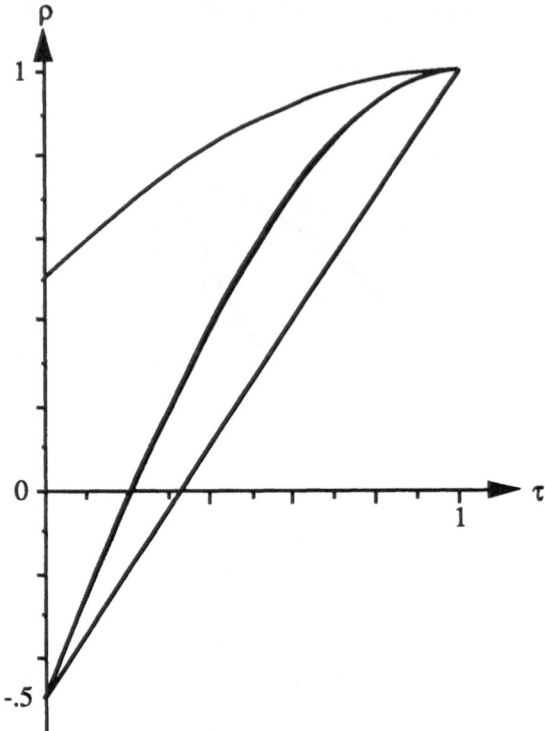

Figure 6. τ and ρ for a family of Sherwood-Taylor copulas.

Members of a family of singular copulas called *shuffles of Min* [Mikusiński, Sherwood, and Taylor (1990, and their contribution to these proceedings)] provide examples for which (τ,ρ) lies on the boundary of the τ,ρ-region. Shuffles of Min are defined informally as follows: "The mass distribution for a shuffle of Min can be obtained by (i) placing the mass distribution for M on I^2, (ii) cutting I^2 vertically into a finite number of strips, (iii) shuffling the strips with perhaps some of them flipped around their vertical axes of symmetry, and then (iv) reassembling them to form the square again." Special cases have been considered by Kimeldorf and Sampson (1978) and Marshall (1989).

Let $\theta \in (0,1]$, set $n = [1/\theta]$ (the greatest integer function), and define

$$C_\theta(u,v) = \begin{cases} \max[(k-1)\theta, u+v-k\theta], & (u,v) \in [(k-1)\theta, k\theta)^2, \ k = 1,2,\dots,n; \\ \max[n\theta, u+v-1], & (u,v) \in [n\theta, 1]^2; \\ \min[u,v], & \text{otherwise.} \end{cases}$$

This copula is a shuffle of Min which assigns mass to the line segments joining the points $((k-1)\theta, k\theta)$ to $(k\theta, (k-1)\theta)$ for $k=1,2,\ldots,n$; and the segment joining $(n\theta, 1)$ to $(1, n\theta)$ (if $n\theta < 1$). Then

$$\tau = 2n\theta(2 - n\theta) - 2n\theta^2 - 1$$

and

$$\rho = 2n\theta^3(n^2 - 1) + 6n\theta(1 - n\theta) - 1.$$

Two special cases are of interest:

Case 1. Let $\theta \in (1/2, 1]$ so that $n = 1$. Then $\tau = 4\theta(1 - \theta) - 1$, $\rho = 6\theta(1 - \theta) - 1$ and hence $\rho = \frac{3}{2}\tau + \frac{1}{2}$. Thus for $\tau \le 0$ every point (τ, ρ) on the upper boundary of the τ, ρ-region is attainable. The copulas in this subfamily are given by

$$C_\theta(u,v) = \begin{cases} \max[0, u + v - \theta], & (u,v) \in [0,\theta)^2; \\ \max[\theta, u + v - 1], & (u,v) \in [\theta,1]^2; \\ \min[u, v], & \text{otherwise.} \end{cases}$$

We obtain the identical expressions for τ and ρ when $\theta \in [0, 1/2)$ [see Marshall (1989)].

Case 2. Let $\theta = 1/n$, $n \in N$. Then $\tau = 1 - 2\theta$ and $\rho = 1 - 2\theta^2$; so that $\rho = \frac{1}{2} + \tau - \frac{1}{2}\tau^2$. Thus for $\tau \ge 0$ the points $(\tau, \rho) = (1 - 2/n, 1 - 2/n^2)$ for $n \in N$ on the upper boundary of the τ, ρ-region are attainable. [For another derivation of the results in these two cases, see Kruskal (1958).] Similar results can be obtained for the lower boundary of the τ, ρ-region by considering members of the family

$$C_\theta(u,v) = \begin{cases} \min[u-(k-1)\theta, v-1+k\theta], & (u,v) \in [(k-1)\theta, k\theta] \times [1-k\theta, 1-(k-1)\theta], \ k=1,2,\ldots,n; \\ \min\{u-n\theta, v\}, & (u,v) \in [n\theta, 1] \times [0, 1-n\theta]; \\ \max[0, u+v-1], & \text{elsewhere.} \end{cases}$$

In Figure 7 we have again graphed the τ, ρ-region, augmented by representations of the copulas attaining certain boundary values. These "representations" of the copulas illustrate the line segments upon which the probability mass is located.

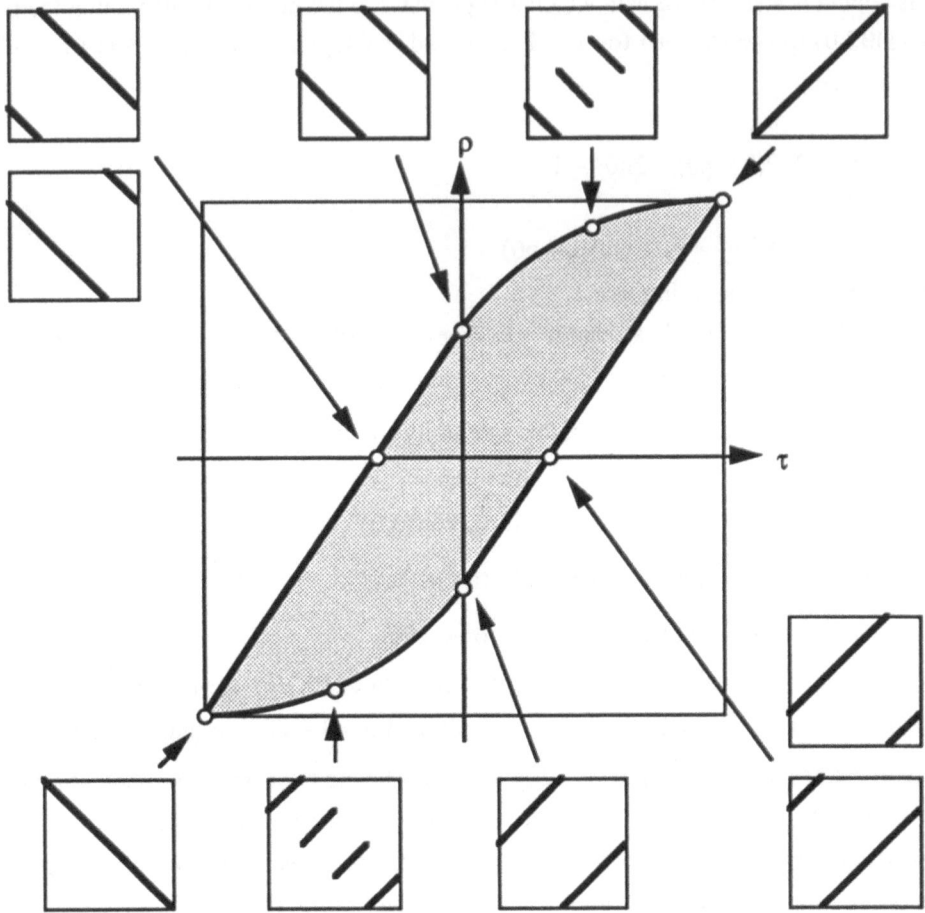

Figure 7. The τ,ρ-region with representations of "boundary" copulas.

As Hutchinson and Lai (1990) note, "... [for a given value of τ] very high ρ only occurs with negative correlation locally contrasted with positive overall correlation, and very low ρ only with negative overall correlation contrasted with positive correlation locally."

We can use shuffles of Min to produce random variables X and Y which are *mutually completely dependent* (Lancaster, 1963), i.e. Pr{Y=g(X)}=1 for some one-to-one function g; but which have any desired value for τ or ρ in [-1,1]. For θ∈ [0,1/2], let C_θ denote the shuffle of Min which distributes the probability mass on three line segments, one joining

(0,1) to (θ,1–θ), the second joining (θ,θ) to (1–θ,1–θ), and third joining (1–θ,θ) to (1,0); i.e.,

$$C_\theta(u,v) = \begin{cases} \min[u, v] - \theta, & (u,v) \in [\theta, 1-\theta]^2; \\ \max[0, u + v - 1], & \text{otherwise.} \end{cases}$$

Then $\tau = 2(1 - 2\theta)^2 - 1$ and $\rho = 2(1 - 2\theta)^3 - 1$; whence τ and ρ may assume all values in [–1,1]. Melnick and Tenenbein (1982) have shown that when the margins are standard normal ($F = G = \Phi$) and $\theta \cong \Phi(-1.54)$, X and Y are mutually completely dependent *uncorrelated* (in the sense of Pearson's correlation coefficient) normal random variables.

3.7. COMMENTS AND OPEN QUESTIONS

We conclude this section with several observations and suggestions for further work.

1. In one sense, τ measures *concordance* (the probability of concordance minus the probability of discordance) while ρ measures *correlation* (between F(X) and G(Y)). While both concordance and correlation are forms of association, they are not the same, and hence we expect different relationships between τ and ρ for different families of copulas. But consider the following four one-parameter ($\theta \in [-1,1]$) families: (a) Copulas for the standard bivariate normal distribution with (Pearson) correlation coefficient θ; (b) the Plackett family, suitably reparametrized [Plackett (1965), Mardia (1967)], (c) the Frank family, suitably reparametrized [Frank (1979), Nelsen (1986), Genest (1987)], and (d) the Raftery family from Section 3.3 above. Each member of each of these families has full support on $(0,1)^2$ (for $-1<\theta<1$) and satisfies the five conditions proposed by Kimeldorf and Sampson (1975): (i) $C_1=M$; (ii) $C_0=\Pi$; (iii) $C_{-1}=W$; (iv) for fixed (u,v), C_θ is continuous in θ on [–1,1]; and (v) for $\theta \in (-1,1)$, $C_\theta(u,v)$ is absolutely continuous. The curves relating ρ to τ are remarkably close to one another for the four families, as shown in Figure 8 (only shown for $\tau \geq 0$) . What are the bounds on ρ as a function of τ for such families?

2. It is well-known that when sampling from a bivariate population in which X and Y are independent (or nearly so), the sample statistic corresponding to ρ is about 50% larger than that corresponding to τ. Figures 2, 3 and 8 suggest a similar behavior for families of copulas which include Π. Specifically, suppose $\{C_\theta(u,v)\}$ is a family of copulas indexed

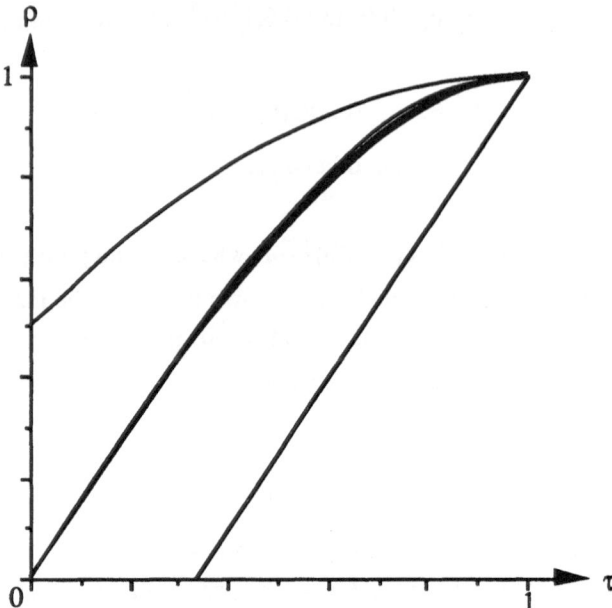

Figure 8. τ and ρ for the standard bivariate normal, Plackett, Frank, and Raftery families.

by the (possibly multidimensional) parameter θ such that $C_{\theta_0}=\Pi$ and C_θ is a continuous function of θ at θ_0. For all such families considered in this section (as well as for a number of others, see Hutchinson and Lai (1990)), we have

$$\lim_{\theta \to \theta_0} \frac{d\rho}{d\tau} = \frac{3}{2}.$$

Does this always hold for such families of copulas?

3. We have seen that some but not all points (τ,ρ) on the boundary of the τ,ρ-region are attainable with shuffles of Min. Are the remaining points attainable? If a point (τ,ρ) on the boundary of the τ,ρ-region is attainable by a copula C, must C be a shuffle of Min as given in section 3.6? [See Kruskal (1958).]

4. For both the generalized Cuadras-Augé family given by (3.2) and the related family given by (3.4) the probability mass on the singular component of a copula is numerically equal to τ. What characterizes all such copulas with this property? Answering this question is equivalent to finding all copulas C which satisfy (when the integrals exist)

$$\iint_{I^2} \frac{\partial^2}{\partial u \partial v} C(u,v) \, du \, dv = 4 \iint_{I^2} \frac{\partial}{\partial u} C(u,v) \frac{\partial}{\partial v} C(u,v) \, du \, dv.$$

4. Some Concepts of Positive Dependence

4.1. DEFINITIONS

In this section we will examine several well-known concepts of positive dependence, and their relationship to the geometry of the surface z=C(u,v). We begin with the definitions of the various concepts, and will use "increasing" and "decreasing" for "nondecreasing" and "nonincreasing," respectively [Lehmann (1966), Esary and Proschan (1972), Barlow and Proschan (1975), Shaked (1977)]:

(i) X and Y are *positive quadrant dependent* if

$\Pr\{X \le x, Y \le y\} \ge \Pr\{X \le x\}\Pr\{Y \le y\}$ for all x and y.

We denote this by PQD(X,Y).

(ii) Y is *left tail decreasing* in X if

$\Pr\{Y \le y \mid X \le x\}$ is decreasing in x for all y.

We will write LTD(Y|X).

(iii) Y is *right tail increasing* in X if

$\Pr\{Y > y \mid X > x\}$ is increasing in x for all y.

We write RTI(Y|X).

(iv) Y is *stochastically increasing* in X if

$\Pr\{Y > y \mid X = x\}$ is increasing in x for all y.

[Lehmann (1966) would say that Y is *positive regression dependent* on X.] We write SI(Y|X). LTD(X|Y), RTI(X|Y) and SI(X|Y) are defined analogously.

Each of these concepts is independent of the marginal distributions of X and Y, and can be expressed in terms of properties of the copula C of X and Y.

Theorem 4.1. Let C denote the copula of the random variables X and Y. Then:

(i)   PQD(X,Y) if and only if $C(u,v) \ge uv$ for all u and v;

(ii) LTD(Y|X) if and only if $C(u,v)/u$ is decreasing in u for all v;

(iii) RTI(Y|X) if and only if $[1 - u - v + C(u,v)]/(1 - u)$ is increasing in u for all v, or equivalently, if $[v - C(u,v)]/(1 - u)$ is decreasing in u for all v; and

(iv) SI(Y|X) if and only if $\frac{\partial}{\partial u}C(u,v)$ is decreasing in u for all v. [Note: If $\frac{\partial}{\partial u}C(u,v)$ does not exist as an ordinary partial derivative, we assume that for each $u \in (0,1)$ it can be defined as an appropriate one-sided partial derivative.]

Proof. Parts (i)-(iii) are immediate. For (iv), we note that (recalling that u = F(x) is increasing in x)

$$Pr\{Y \leq y|X=x\} = \frac{Pr\{Y \leq y, X=x\}}{Pr\{X=x\}} = \frac{\frac{\partial}{\partial x}H(x,y)}{\frac{d}{dx}F(x)} = \frac{\frac{\partial}{\partial u}C(u,v)\frac{\partial u}{\partial x}}{\frac{\partial u}{\partial x}} = \frac{\partial}{\partial u}C(u,v) \ .$$

Corollary 4.2. Under the hypotheses of Theorem 4.1, we have

(i) LTD(Y|X) if and only if $\frac{\partial}{\partial u}C(u,v) \leq C(u,v)/u$; and

(ii) RTI(Y|X) if and only if $\frac{\partial}{\partial u}C(u,v) \geq [v - C(u,v)]/(1 - u)$.

The concepts of LTD and RTI are easily related via the survival copula (see Section 3.2). Let us write RTI(Y|X;C_H) to indicate that Y is RTI in X where the bivariate distribution function for X and Y is $C_H(F(x),G(y))$, and similarly for LTD.

Corollary 4.3. RTI(Y|X;C_H) if and only if LTD(Y|X;$C_{\overline{H}}$); and LTD(Y|X;C_H) if and only if RTI(Y|X;$C_{\overline{H}}$).

Proof. From Theorem 4.1(iii), RTI(Y|X;C_H) if and only if $[1-u-v+C_H(u,v)]/(1-u)$ is increasing in u for all v. Replacing u by 1-u and v by 1-v yields $[u+v-1+C_H(1-u,1-v)]/u$ decreasing in u for all v, or equivalently, $C_{\overline{H}}(u,v)/u$ decreasing in u for all v; hence LTD(Y|X;$C_{\overline{H}}$). The second part of the corollary is immediate since $\overline{\overline{H}}=H$.

4.2. GEOMETRIC INTERPRETATIONS

We now relate these concepts to geometric properties of C. We first note that the geometric interpretation of positive quadrant dependence is obvious—the surface z = C(u,v) lies on or above the surface z = Π(u,v) = uv.

Theorem 4.4. Y is stochastically increasing in X if and only if C(u,v) is a concave function of u for every v.

Proof. This theorem follows from the observations that a function f:(0,1)→R is concave if and only if its one-sided derivatives f'_+ and f'_- are decreasing [Roberts and Varberg (1973, Section 11, Theorem B and Remark E)]; and that for all x∈ (0,1), $f'_-(x) \geq f'_+(x)$.

For the next two theorems, the notion of a starlike region is useful. We say that a region S is *starlike with respect to the point P in S* if for every point Q in S, all points on the line segment PQ are points of S. We will be concerned with regions of the form

$S_v = \{(u,z)| u \in [0,1], 0 \le z \le C(u,v)\}$. Geometrically, S_v is the set of points in the intersection of the solid $0 \le z \le C(u,v)$, $(u,v) \in I^2$ with a plane perpendicular to the v-axis. Figure 9 illustrates such a region for the copula in Fréchet's family given by (3.1) with $\alpha=1/2$, $\beta=0$.

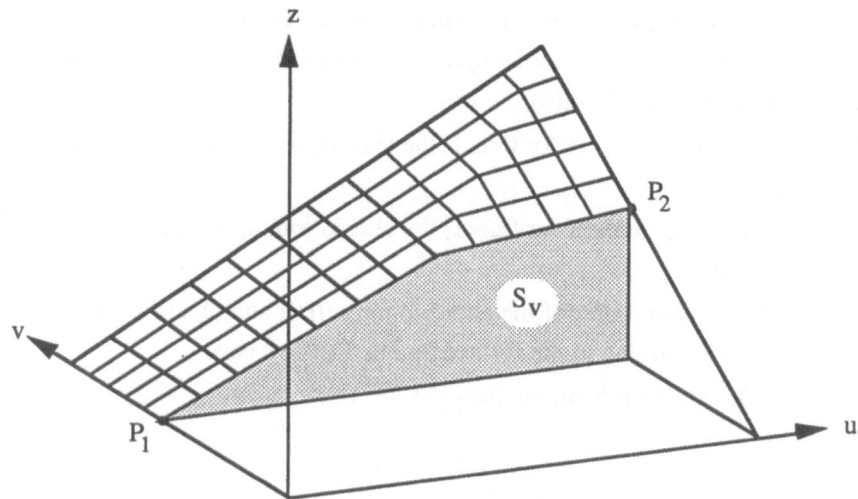

Figure 9. Example of a star-like region.

Theorem 4.5. Y is left tail decreasing in X if and only if for every v the region S_v is starlike with respect to the point $P_1=(0,0) \in S_v$.

Proof. Assume LTD(Y|X), fix v, and let $c(u)=C(u,v)$. To show that S_v is starlike with respect to $P_1 \in S_v$, we need only show that for $0 < t < 1$, chords joining $(0,0)$ to $(t,c(t))$ lie below $z = c(u)$. Consider the points t and λt where $0 < \lambda < 1$. Since $\lambda t < t$, $C(\lambda t,v)/\lambda t \ge C(t,v)/t$ (since $C(u,v)/u$ is decreasing in u), or equivalently, $c(\lambda t) \ge \lambda c(t)$. Hence $c(\lambda t + (1-\lambda)0) \ge \lambda c(t) + (1-\lambda)c(0)$, so that every point on the chord joining $(0,0)$ to $(t,c(t))$ lies below $z=c(u)$, and thus S_v is starlike with respect to P_1.

Conversely, assume that S_v is starlike with respect to $P_1 \in S_v$. Let $0 \le u_1 < u_2 \le 1$. Since the chord joining $(0,0)$ to $(u_2,c(u_2))$ lies below $z=c(u)$, we have

$$c(u_1) = c\left(\frac{u_1}{u_2}u_2 + (1 - \frac{u_1}{u_2})0\right) \ge \frac{u_1}{u_2}c(u_2) + (1 - \frac{u_1}{u_2})c(0) = \frac{u_1}{u_2}c(u_2).$$

and hence $C(u_1,v)/u_1 \ge C(u_2,v)/u_2$. Thus LTD(Y|X).

Theorem 4.6. Y is right tail increasing in X if and only if for every v the region S_v is starlike with respect to the point $P_2=(1,v)\in S_v$.

Proof. The proof is similar to that of the preceding theorem, and therefore omitted.

The geometric interpretation of the four concepts can be summarized as follows. Consider the curve $z=c(u)$ in the plane determined by S_v. Then

(i) SI(Y|X) if and only if for $0\le u_1<u_2\le 1$, the chords joining two points $(u_1,c(u_1))$ and $(u_2,c(u_2))$ lie below the curve $z=c(u)$;

(ii) LTD(Y|X) if and only if for $0<t\le 1$, the chords joining $(0,0)$ and $(t,c(t))$ lie below the curve $z=c(u)$;

(iii) RTI(Y|X) if and only if for $0\le t<1$; the chords joining $(t,c(t))$ and $(1,v)$ lie below the curve $z=c(u)$; and

(iv) PQD (Y|X) if and only if the chord joining $(0,0)$ and $(1,v)$ lies below the curve $z=c(u)$. This chord is a ruling of the surface for Π. Thus we have

Theorem 4.7. The following implications are valid:

$$
\begin{array}{ccc}
 & \text{LTD(Y|X)} & \\
 & \nearrow \qquad \searrow & \\
\text{SI(Y|X)} & & \text{PQD(X,Y)} \\
 & \searrow \qquad \nearrow & \\
 & \text{RTI(Y|X)} &
\end{array}
$$

Other proofs of Theorem 4.7 are given by Barlow and Proschan (1975) and Shaked (1977).

4.3. COMMENTS AND OPEN QUESTIONS

We conclude this section with some observations and suggestions for further work.

1. Do other positive dependence concepts have similiar geometric interpretations? For example, X and Y are *right corner set increasing* [RCSI(X,Y)] if $Pr\{X>x,Y>y|X>x',Y>y'\}$ is increasing in x' and y' for all x and y [Harris (1970)]. But as Shaked (1977) shows, this is equivalent to

$$\frac{Pr\{X>x,Y>y\}}{Pr\{X>x,Y>y'\}} \ge \frac{Pr\{X>x',Y>y\}}{Pr\{X>x',Y>y'\}}$$

for all $x<x'$, $y<y'$. Since RCSI(X,Y) implies both RTI(Y|X) and RTI(X|Y) [set $y=-\infty$ or $x=-\infty$ above], a geometric property of C equivalent to RCSI should not be difficult to find.

2. What is the relationship between these concepts of positive dependence and the correlation coefficients discussed earlier? It is well-known [Lehmann (1966)] that PQD(X,Y) implies that $\rho \geq 0$ and $\tau \geq 0$. Based on the examination of many families of copulas, Hutchinson and Lai (1990) hypothesize that if SI(Y|X) and SI(X|Y), then

$$-1 + \sqrt{1 + 3\tau} \leq \rho \leq \min[\frac{3\tau}{2}, 2\tau - \tau^2].$$

The results of Theorem 4.1 with the expressions for τ and ρ from (2.3 and 2.7) suggest that it should not be difficult to obtain bounds for ρ in terms of τ for several of the positive dependence concepts.

Acknowledgment

The author would like to express his gratitude to G. A. Fredricks, B. Schweizer, and R. M. Tardiff for their valuable assistance with several sections in this paper; and to Mount Holyoke College in South Hadley, Massachusetts, where he was a Visiting Professor in the Department of Mathematics while completing work on this paper.

References

Barlow, R. E. and Proschan, F. (1975) *Statistical Theory of Reliability and Life Testing: Probability Models,* Holt, Rinehart and Winston, Inc., New York.

Conway, D. A. (1979) Multivariate distributions with specified marginals, Technical Report No. 145, Department of Statistics, Stanford University.

Cuadras, C. M. and Augé, J. (1981) 'A continuous general multivariate distribution and its properties', *Commun. Statist.- Theor. Meth.,* **A10**, 339-353.

Daniels, H. E. (1950) 'Rank correlation and population models', *J. Roy. Statist. Soc. Ser. B.,* **12**, 171-181.

Devroye, L. (1986) *Non-Uniform Random Variate Generation,* Springer-Verlag, New York.

Durbin, J. and Stuart, A. (1951) 'Inversions and rank correlation coefficients', *J. Roy. Statist. Soc. Ser. B.,* **2**, 303-309.

Esary, J. D. and Proschan, F. (1972) 'Relationships among some concepts of bivariate dependence', *Ann. Math. Statist.,* **43**, 651-655.

Frank, M. J. (1979) 'On the simultaneous associativity of F(x,y) and x + y − F(x,y)', *Aequationes Math.,* **19**, 194-226.

Fréchet, M. (1958) 'Remarques au sujet de la note précédente', *C. R. Acad. Sci. Paris..* **246**, 2719-2720.

Genest, C. (1987) 'Frank's family of bivariate distributions' *Biometrika,* **74**, 549-555.

Harris, R. (1970) 'A multivariate definition for increasing hazard rate distribution functions', *Ann. Math. Statist.,* **41**, 713-717.

Hutchinson, P. and Lai, C. D. (1990) *Continuous Bivariate Distributions, Emphasizing Applications,* Rumsby Scientific Publishing, Adelaide.

Kamiński, A., Mikusiński, P., Sherwood, H. and Taylor, M. D. (1988) 'Properties of a special class of doubly stochastic measures', *Aequationes Math.*, **36**, 212-229.

Kendall, M. G. (1962) *Rank Correlation Methods,* Charles Griffin and Co., London.

Kimeldorf, G. and Sampson, A. R. (1978) 'Monotone dependence', *Ann. Statist.*, **6**, 895-903.

Kruskal, W. H. (1958) 'Ordinal measures of association', *J. Amer. Statist. Assoc.*, **53**, 814-861.

Lancaster, H. O. (1963) 'Correlation and complete dependence of random variables', *Ann. Math. Statist.*, **34**, 1315-1321.

Lehmann, E. L. (1966) 'Some concepts of dependence', *Ann. Math. Statist.*, **37**, 1137-1153.

Mardia, K. V. (1967) 'Some contributions to contingency-type bivariate distributions', *Biometrika*, **54**, 235-249.

Mardia, K. V. (1970) *Families of Bivariate Distributions,* Charles Griffin and Co., London.

Marshall, A. W. (1989) 'A bivariate uniform distribution,' in L. J. Gleser, M. D. Perlman, S. J. Press and A. R. Sampson (eds.), *Contributions to Probability and Statistics,* 99-106, Springer-Verlag, New York.

Marshall, A. W. and Olkin, I. (1967) 'A multivariate exponential distribution', *J. Amer. Statist. Assoc.*, **62**, 30-44.

Melnick, E. L. and Tenenbein, A. (1982) 'Misspecifications of the normal distribution', *Amer. Statist.*, **36**, 372-373.

Mikusiński, P., Sherwood, H. and Taylor, M. D. (1990) Shuffles of Min, To appear.

Muliere, P. and Scarsini, M. (1987) 'Characterization of a Marshall-Olkin type class of distributions', *Ann. Inst. Statist. Math.*, **39**, 429-441.

Nelsen. R. B. (1986) 'Properties of a one-parameter family of bivariate distributions with specified marginals', *Commun. Statist.-Theory Meth.*, **15**, 3277-3285.

Plackett, R. L. (1965) 'A class of bivariate distributions', *J. Amer. Statist Assoc.*, **60**, 516-522.

Raftery, A. E. (1984) 'A continuous multivariate exponential distribution', *Commun. Statist.-Theor. Meth.*, **13**, 947-965.

Raftery, A. E. (1985) 'Some properties of a new continuous bivariate exponential distribution', *Statistics & Decisions,* Supp. Issue No. 2, 53-58.

Roberts, A. W. and Varberg, D. E. (1973) *Convex Functions,* Academic Press, New York.

Schweizer, B. and Sklar, A. (1983) *Probabilistic Metric Spaces,* Elsevier Science Publishing Co., New York.

Schweizer, B. and Wolff, E. F. (1981) 'On nonparametric measures of dependence for random variables', *Ann. Statist.*, **9**, 879-885.

Shaked, M. (1977) 'A family of concepts of dependence for bivariate distributions', *J. Amer. Statist. Assoc.*, **72**, 642-650.

Sherwood, H. and Taylor, M. D. (1988) 'Doubly stochastic measures with hairpin support', *Probab. Th. Rel. Fields.*, **78**, 617-626.

Tchen, A. (1980) 'Inequalities for distributions with given marginals', *Ann. Probab.*, **8**, 814-827.

Whitt, W. (1976) 'Bivariate distributions with given marginals', *Ann. Statist.*, **4**, 1280-1289.

CONVOLUTIONS FOR DEPENDENT RANDOM VARIABLES

M. J. FRANK
Department of Mathematics
Illinois Institute of Technology
Chicago, Illinois 60616   U.S.A.

ABSTRACT.   A two-parameter family $\sigma_{C,L}$ of binary operations arises naturally as the distributional counterpart of operations on random variables.   Here L is an arbitrary measurable function of two r.v.'s and C is their copula, or dependence function. Ordinary convolution is the particular member $\sigma_{Prod,sum}$, i.e., when L is addition and C is the independence copula.   In this paper, we investigate the algebraic and analytic structure of the $\sigma_{C,L}$, extending earlier results obtained for subfamilies.

1. Copulas and the Operations $\sigma_{C,L}$

Convolution maps the probability distributions of two independent random variables to the distribution of their sum.   This binary operation can be extended in two directions: (1) replace addition by other operations on random variables; (2) allow stochastic relations other than independence.

 The purpose of this paper is to present a unified study of the binary operations on one-dimensional distribution functions which are induced by operations on real-valued random variables.   Our development rests on isolating the dependence structure from the individual distributions through the notion of a copula, introduced by A. Sklar in [13].

 A (two-dimensional) *copula* is a function C: $[0,1]^2 \rightarrow [0,1]$ that satisfies the boundary conditions

$$C(s,0) = C(0,s) = 0, \qquad C(s,1) = C(1,s) = s, \qquad (1.1)$$

and the monotonicity condition

$$C(s_1,t_1) - C(s_2,t_1) - C(s_1,t_2) + C(s_2,t_2) \geq 0 \qquad (1.2)$$

75

whenever $s_1 \leq s_2$ and $t_1 \leq t_2$.

Let $\mathcal{C}$ be the set of copulas. It is easily verified that each C in $\mathcal{C}$ is continuous on $[0,1]^2$, is non-decreasing in each place, and satisfies the pointwise inequalities

$$W \leq C \leq Min, \qquad (1.3)$$

where W and Min are the copulas given by

$$W(s,t) = \text{maximum } (s+t-1,0),$$

$$Min(s,t) = \text{minimum } (s,t).$$

Copulas link joint distributions to their margins: Let K be a bivariate distribution function with margins F and G. Then there is a copula C such that

$$K(u,v) = C(F(u),G(v)) . \qquad (1.4)$$

In the other direction, given C in $\mathcal{C}$ and distributions F and G, the function K defined by (1.4) is a bivarate distribution with margins F and G.

In particular, if X and Y are extended real-valued random variables (r.v.'s), defined on a common probability space, with individual distributions F_X and F_Y and joint distribution $F_{X,Y}$, then there is a $C_{X,Y}$ in $\mathcal{C}$ such that

$$F_{X,Y}(u,v) = C_{X,Y}(F_X(u),F_Y(v)) . \qquad (1.5)$$

Note that if F_X and F_Y are continuous, $C_{X,Y}$ is unique. The copula in (1.5) is said to *connect* X and Y. We shall refer to $C_{X,Y}$ as a copula of X and Y.

The copula of two r.v.'s thus reveals their dependence structure. For instance, X and Y are independent precisely when they are connected by the copula Prod, where

$$\text{Prod } (s,t) = st .$$

Stochastic interpretations of W and Min are well-known: continuous X and Y are connected by Min (resp., W) if and only if Y is a.s. an increasing (resp., decreasing) function of X. Observe that (1.3) and (1.4) combine to yield the Fréchet bounds.

Copulas provide a natural framework for the study of distributions with fixed margins. For a summary of results and a guide to the extensive literature, consult [12] and B. Schweizer's contribution to this volume.

In order to define the operations treated in this paper, we begin with some notation. Let I be a closed subinterval of the extended real line, i.e., I = [a,b], with $-\infty \leq a < b \leq \infty$. Denote by Δ_I the set of distribution functions whose support is contained in I, i.e., non-decreasing F for which $F(x) = 0$ when $x < a$ and $F(x) = 1$ when $x > b$. The elements of Δ_I are not assumed to be normalized by left- or right-continuity. Denote by $\mathcal{B}_I$ the set of Borel- measurable two-place functions from I^2 to I. The elements of $\mathcal{B}_I$ may be viewed as binary operations on I.

Now for any r.v.'s X,Y with F_X, F_Y in Δ_I and any L in $\mathcal{B}_I$, the r.v. L(X,Y) is well-defined and $F_{L(X,Y)}$ also belongs to Δ_I. Moreover, in view of (1.5), for every real x,

$$F_{L(X,Y)}(x) = \iint_{L(u,v)<x} dC_{X,Y}(F_X(u), F_Y(v)) .$$ (1.6)

This immediately justifies the definition of the operations which are the subject of this paper.

DEFINITION 1.1. Let C belong to $\mathcal{C}$ and L to $\mathcal{B}_I$. The mapping $\sigma_{C,L}$ is the binary operation on Δ_I whose value at (F,G) in Δ_I is the function $\sigma_{C,L}(F,G)$ in Δ_I given by

$$\sigma_{C,L}(F,G)(-\infty) = 0, \qquad \sigma_{C,L}(F,G)(\infty) = 1,$$

and for $-\infty < x < \infty$,

$$\sigma_{C,L}(F,G)(x) = \iint_{L(u,v)<x} dC(F(u), G(v)) .$$ (1.7)

Thus, by virtue of (1.6), if C connects X and Y,

$$F_{L(X,Y)} = \sigma_{C,L}(F_X, F_Y) .$$ (1.8)

In the subsequent study of the two-parameter family of operations $\sigma_{C,L}$, the most important underlying intervals I are $[-\infty,\infty]$ and $[0,\infty]$. For simplicity of notation we shall write

$$\Delta = \Delta_{[-\infty,\infty]}, \qquad \mathcal{B} = \mathcal{B}_{[-\infty,\infty]}$$

and

$$\Delta_+ = \Delta_{[0,\infty]}, \qquad \mathcal{B}_+ = \mathcal{B}_{[0,\infty]} .$$

Let $\mathcal{D}$ be the subset of "non-defective" elements of Δ, i.e., distributions whose support is contained in $(-\infty,\infty)$, or equivalently, of finite-valued r.v.'s. Since we insist that an element L of $\mathcal{B}$ take on finite values at finite arguments, it follows from (1.7) and (1.6) that $\mathcal{D}$ is closed under $\sigma_{C,L}$, and an identical statement holds for L in $\mathcal{B}_+$ and $\mathcal{D}_+$.

Note that ordinary convolution $*$ on Δ or Δ_+ is the member of the family (1.7) for which L is addition (in $\mathcal{B}$ or $\mathcal{B}_+$) and C is the copula of independence, i.e.,

$$F*G = \sigma_{\text{Prod,sum}}(F,G) .$$

Our investigation will focus on the algebraic and analytic structure of the family $\sigma_{C,L}$, especially with regard to desirable properties enjoyed by convolution. Thus we seek to determine which of the $\sigma_{C,L}$ are order-preserving, associative, linear, homogeneous, continuous, and so forth. It turns out that $\sigma_{C,L}$ inherits many of these properties from the corresponding properties of C or L or both, while others are more complicated. We shall also point out some of the implications of our results in the theory of distributions with fixed margins.

The development presented here may be viewed as a continuation of research begun during the 1970's by the author in two papers [5,6]. In the first of these [5], the principal result is the characterization of associativity in the family $\sigma_{C,\text{sum}}$ via reduction to and solution of certain real functional equations. The second paper [6] deals with some related equations (which are also of independent interest) and preliminary attempts at extension to more general L. We shall frequently refer to the results in these articles, and especially to several of the proofs.

In Section 2, we explore some simple consequences of the definition of $\sigma_{C,L}$ and give some elementary examples. Sections 3 and 4 are devoted primarily to the question of associativity among the $\sigma_{C,L}$: in Section 3, we obtain necessary conditions which lead to an important family of copulas; in Section 4, we complete the discussion. Section 5 treats linearity, homogeneity, and continuity, relates our results to the generalized convolutions of K. Urbanik, and touches on several unsolved problems.

2. Elementary Properties and Examples

The Lebesgue-Stieltjes integrals (1.7) which define the operations $\sigma_{C,L}$ are, except for very special C and L, difficult to handle. In particular, since the boundary of the planar region $L(u,v) < x$ is a usually curve, evaluation of the measure induced by C, F, and G is often tricky at best.

When L is either of the operations max or min on any interval I, however, these regions are rectangular, the measures are easily evaluated, and the $\sigma_{C,L}$ reduce to simple pointwise operations:

For every C in $\mathscr{C}$ and every F,G in Δ_I ,

$$\sigma_{C,max}(F,G)(x) = C(F(x),G(x)) , \qquad (2.1)$$

and

$$\sigma_{C,min}(F,G)(x) = \overline{C}(F(x),G(x)) , \qquad (2.2)$$

where $\overline{C}$ is the *dual copula* of C, given by

$$\overline{C}(s,t) = s + t - C(s,t) . \qquad (2.3)$$

Our principal technique for revealing the structure of the $\sigma_{C,L}$ is based on the following observation: by the judicious choice of simple F and G, one can frequently translate or reduce properties of $\sigma_{C,L}$ to properties of C and L, either by evaluating the integral $\sigma_{C,L}(F,G)$ directly or by appealing to its stochastic interpretation (1.8).

As an illustration, for u in $[-\infty,\infty]$ let ϵ_u be the d.f. of the constant r.v. X = u, i.e., the unit step function at u. Then it follows immediately from (1.8) that for each I containing u, each L in $\mathscr{B}_I$ and F_Y in Δ_I ,

$$\sigma_{C,L}(\epsilon_u,F_Y) = F_{L(u,Y)} . \qquad (2.4)$$

In particular, $\sigma_{C,L}(\epsilon_u,F_Y) = F_Y$ if and only if L(u,Y) = Y a.s., whence we conclude:

THEOREM 2.1. For C in $\mathscr{C}$, L in $\mathscr{B}_I$, and u in I, ϵ_u is an identity (resp., a null element) for the operation $\sigma_{C,L}$ on Δ_I if and only if u is an identity (resp., a null element) of L.

Letting Y = v in (2.4), we obtain

$$\sigma_{C,L}(\epsilon_u,\epsilon_v) = \epsilon_{L(u,v)} \text{ for all u,v in I.} \qquad (2.5)$$

An immediate consequence of (2.5) and (1.7) is the following:

THEOREM 2.2. If the operation $\sigma_{C,L}$ is commutative, then so is L. Moreover, if both C and L are commutative, then $\sigma_{C,L}$ is commutative.

And repeated use of (2.5) yields :

THEOREM 2.3. If $\sigma_{C,L}$ is associative, L must be associative.

Recall that a binary operation ϕ on a set S is associative if it is a solution of the functional equation

$$\phi(\phi(x,y),z) = \phi(x,\phi(y,z)), \qquad x,y,z \ \ in \ \ S.$$

To this point we have not imposed the usual normalization of Δ which requires its elements to be, say, left-continuous. In fact, it is easy to find operations L for which $\sigma_{C,L}$ fails to preserve any such normalization. Since d.f.'s are actually equivalence classes, none of our results are materially affected; yet it is often convenient in calculating integrals to deal with normalized d.f.'s. The proof of Theorem 5 of [5] for $\sigma_{C,\text{sum}}$ can be adapted to yield:

THEOREM 2.4. Let L in $\mathcal{B}_I$ be non-decreasing in each place on I^2. If F,G in Δ_I are left-continuous on $(-\infty,\infty)$, then so is $\sigma_{C,L}(F,G)$.

Henceforth, whenever L is non-decreasing, we shall impose left-continuity on elements of Δ_I.

THEOREM 2.5. Let C in $\mathcal{C}$ and L in $\mathcal{B}_I$ be given. Then $\sigma_{C,L}$ preserves the pointwise partial order on Δ_I, i.e.,

$$\sigma_{C,L}(F_1,G_1) \le \sigma_{C,L}(F_2,G_2) \ \ \text{whenever} \ F_1 \le F_2, \ G_1 \le G_2 ,$$

if and only if L is non-decreasing in each place on I^2.

Proof. In one direction, the result is immediate. Suppose that L fails to be non-decreasing; choose u,v,w in I with $u < v$ and, say, $L(u,w) > L(v,w)$. Then $\epsilon_u > \epsilon_v$, but by (2.5)

$$\sigma_{C,L}(\epsilon_u,\epsilon_w) = \epsilon_{L(u,w)} < \epsilon_{L(v,w)} = \sigma_{C,L}(\epsilon_v,\epsilon_w) .$$

In the other direction, fix $F_1 \le F_2$ in Δ_I, G in Δ_I, and x in I. Let $L_x = \{(u,v) : L(u,v) < x\}$, and for $k = 1,2$, let D_k denote the region of $[0,1]^2$ given by

$$D_k = \{(s,t) : s \le F_k(u), \ t \le G(v) \ \text{for some} \ (u,v) \ \text{in} \ L_x\}.$$

Since L is non-decreasing, it follows that $\sigma_{C,L}(F_k,G)(x)$ is equal to the value of the C-measure of D_k. But $F_1 \le F_2$ obviously implies that $D_1 \subseteq D_2$, and so

$$\sigma_{C,L}(F_1,G)(x) \le \sigma_{C,L}(F_2,G)(x) .$$

Repeating the argument for F and $G_1 \le G_2$ completes the proof.

3. Associativity of $\sigma_{C,L}$: Necessary Conditions

We now address the question of associativity among the operations $\sigma_{C,L}$; that is, we wish to solve the equation

$$\sigma_{C,L}(\sigma_{C,L}(F,G),H) = \sigma_{C,L}(F,\sigma_{C,L}(G,H)) \tag{3.1}$$

on Δ_I for L in $\mathcal{B}_I$ and C in $\mathcal{C}$. The required analysis is quite intricate and the answer surprising: except for simple choices of L, associativity of $\sigma_{C,L}$ imposes severe restrictions on the copula C.

It might appear at first glance that as long as L is associative, so is $\sigma_{C,L}$ because for any X,Y,Z,

$$F_{L(L(X,Y),Z)} = F_{L(X,L(Y,Z))} \; .$$

However, the r.v.'s here are rarely connected by the same copula C. Thus, in a sense, associativity of $\sigma_{C,L}$ also demands the preservation of the dependence relation. And in particular $\sigma_{C,L}$ does not inherit associativity from that of C.

We mention a trivial exception. Let $L(u,v) = u$. Then for every F and G, $\sigma_{C,L}(F,G) = F$, whence $\sigma_{C,L}$ is associative for every C in $\mathcal{C}$.

DEFINITION 3.1. Let $\mathcal{L}_I$ be the set of L in $\mathcal{B}_I$ that are non-decreasing in each place, associative, commutative, and continuous on $I^2 = [a,b]^2$ except possibly at the points (a,b) and (b,a).

We shall generally confine our discussion to L in $\mathcal{L}_I$. In view of Theorem 2.3, associativity of L is essential. Many of our results require at least some weak variant of monotonicity and some require continuity. The exceptional points are allowed so as to include important examples such as L = sum in $\mathcal{L}$ and L = multiplication in $\mathcal{L}_+$. Commutativity is a convenience. It will be clear from the proofs which conditions can be relaxed.

EXAMPLES 3.2. For any I, L = max and L = min belong to $\mathcal{L}_I$. By virtue of (2.1) and (2.2),

$$\sigma_{C,max} \text{ is associative iff } C \text{ is associative,}$$

and

$$\sigma_{C,min} \text{ is associative iff } \overline{C} \text{ is associative.}$$

Now define L_0 in $\mathcal{L}$ via

$$L_0(u,v) = u + v - \max(u,v,0) - \min(u,v,0)$$

$$= \begin{cases} \max(u,v), & u,v \le 0, \\ \min(u,v), & u,v \ge 0, \\ 0, & \text{otherwise.} \end{cases} \tag{3.2}$$

Then

$$\sigma_{C,L_0}(F,G)(x) = \begin{cases} C(F(x),G(x)), & x \le 0, \\ \overline{C}(F(x),G(x)), & x > 0. \end{cases}$$

Continuing calculation of the integrals appearing in equation (3.1), we readily obtain:

σ_{C,L_0} is associative iff *both* C and $\overline{C}$ are associative,

i.e., iff

$$C(C(r,s),t) = C(r,C(s,t)) \tag{3.3}$$

and

$$\overline{C}(\overline{C}(r,s),t) = \overline{C}(r,\overline{C}(s,t)) . \tag{3.4}$$

The well-known representation theorems for associative real functions yield all solutions of (3.3) and (3.4) separately, and there are many. Briefly, those associative functions not having interior idempotent elements admit a simple representation via one-place functions. The rest are ordinal sums of these. (The reader is referred to [12] for a thorough discussion and, in particular, for the ordinal sum construction.) But, as we shall see after the next theorem, the two equations taken together have very few solutions.

In [5,6], it is proved that $\sigma_{C,\text{sum}}$ is associative if and only if C is associative and a solution of the equation

$$C(r + s - C(r,s), t) = C(r,t) + C(s,t) - C(C(r,s),t) . \tag{3.5}$$

We now derive these as <u>necessary</u> conditions for the associativity of $\sigma_{C,L}$ for nearly all L. For associative L, write L(u,v,w) in place of L(L(u,v),w).

THEOREM 3.3. Suppose that L in $\mathscr{L}_I$ satisfies the following condition: there exist u < v < w in I such that

$$L(u,u,u) < L(u,u,v) < \min[L(u,v,v),L(u,u,w)] . \tag{3.6}$$

Then in order that $\sigma_{C,L}$ be associative, C must be a solution of equations (3.3) and (3.5).

Proof. We calculate the integrals in (3.1) for appropriate choices of d.f.'s with two-point support, i.e., of the form

$$F_{uv}{}^s(x) = \begin{cases} 0, & x \le u, \\ s, & u < x \le v, \\ 1, & v < x. \end{cases} \qquad (3.7)$$

Set $F = F_{uv}{}^r$, $G = F_{uv}{}^s$, and $H = F_{uw}{}^t$. It follows from (3.6) that $L(u,u) < L(u,v)$ and $L(u,v) < L(v,v)$, whence

$$\sigma_{C,L}(F,G) = \begin{cases} 0, & x < L(u,u), \\ C(r,s), & L(u,u) < x \le L(u,v), \\ r + s - C(r,s), & L(u,v) < x \le L(v,v), \\ 1, & L(v,v) < x. \end{cases}$$

Continuing in this way, equation (3.1) ultimately yields equation (3.3) on $L(u,u,u) < x < L(u,u,v)$ and equation (3.5) on the other interval in (3.6). Since r,s, and t are arbitrary elements of [0,1], the proof is complete.

It is shown in [5,6] that (3.3) and (3.5) together imply (3.4), but that (3.3) and (3.4) do not imply (3.5). Observe that none of the L's in Examples 3.2 satisfy condition (3.6). Note also that continuity of L is not used in the proof of Theorem 3.3.

In solving the system of equations (3.3) and (3.4), the author was led to discover a one-parameter family of copulas of independent interest. For $-\infty \le \alpha \le \infty$, define C_α on $[0,1]^2$ as follows: for $-\infty < \alpha < \infty$, $\alpha \ne 1$, let

$$C_\alpha(s,t) = -\frac{1}{\alpha}\log\left[1 + \frac{(e^{-\alpha s}-1)(e^{-\alpha t}-1)}{e^{-\alpha}-1}\right]; \qquad (3.8a)$$

for $\alpha = -\infty, 0, \infty$, let C_α be the pointwise limits of (3.8), which turn out to be

$$C_{-\infty} = W, \quad C_0 = Prod, \quad C_\infty = Min. \qquad (3.8b)$$

The C_α are a family of associative copulas, C_α is strictly increasing in each place on $[0,1]^2$ for $-\infty < \alpha < \infty$, and the C_α are continuously ordered in the sense that

$$C_\alpha < C_\beta \quad \text{iff} \quad \alpha < \beta,$$

$\lim_{\alpha \to \beta} C_\alpha = C_\beta$ for each β.

Moreover, this is the only family found to date which enjoys all of the above properties and contains the copulas W, Prod, and Min. The C_α, as a consequence, have recently played a prominent role in modeling via distributions with fixed margins. In particular, various properties have been extensively explored by C. Genest [9] and R.B. Nelsen [11]; see also their contributions to this volume.

The copulas (3.8) are in essence the only simultaneous solutions of (3.3) and (3.4):

Let C belong to $\mathscr{C}$ and let $\overline{C}$ be its dual, given by (2.3). Then both C and $\overline{C}$ are associative if and only if

(1) C is a member of the family C_α, $-\infty \le \alpha \le \infty$, in (3.8), or
(2) C is an ordinal sum of members of this family.

(The proof of this result [6] is very complicated, but ultimately involves reduction to a simple differential equation whose solutions generate the family C_α.)

Dual copulas are closely related to survival copulas -- dependence functions that connect joint survivals to their marginal survivals. For C in $\mathscr{C}$, the survival copula C′ is given by

$$C'(s,t) = s + t - 1 + C(1-s,1-t) .$$

Let S_X, S_Y, $S_{X,Y}$ be the survival functions and joint survival functions of X and Y. From their definitions and (1.5), an easy calculation produces

$$S_{X,Y}(u,v) = C_{X,Y}'(S_X(u),S_Y(u)) .$$

It is straightforward to show that C′ is a copula and that C′ is associative precisely when $\overline{C}$ is associative. Thus, it is not surprising that the associativity of $\sigma_{C,L}$ should require associativity of both C and C′. And the theorem quoted in the preceding paragraph exhibits all such copulas. Moreover, it is worth noting that for the family C_α in (3.8), $C_\alpha = C_\alpha'$.

To complete this section, it remains to find the solutions C to the conclusion of Theorem 3.3, i.e., to solve the system (3.3) and (3.5). This was done is [6] by showing directly that, among members of the family (3.8), only Prod and Min satisfy equation (3.5). Thus we have:

THEOREM 3.4. Let L be as in Theorem 3.3. Then if $\sigma_{C,L}$ is associative, either

C = Min, C = Prod, or C is an ordinal sum of Prod.

Theorem 3.4 is a negative result with regard to the search for associative operations on Δ; however, it may be viewed as positive in the sense that it provides a characterization of independence and monotone increasing dependence.

4. Associativity of $\sigma_{C,L}$: Sufficient Conditions

In view of Theorems 3.3 and 3.4, the search for associative $\sigma_{C,L}$ is, apart from simple cases such as in Examples 3.2, confined to C = Min, C = Prod, and the ordinal sums.

We begin with C = Min. The proof of Theorem 7 in [5] for $\sigma_{Min,sum}$ can be extended to establish:

THEOREM 4.1. Suppose that L in $\mathcal{L}_I$ has range I and satisfies

$$L(u_1,v_1) < L(u_2,v_2) \quad \text{whenever } u_1 < u_2, \ v_1 < v_2. \tag{4.1}$$

Then

$$\sigma_{Min,L} = \tau_{Min,L} , \tag{4.2}$$

where $\tau_{Min,L}$ is the binary operation on Δ_I given by

$$\tau_{Min,L}(F,G)(x) = Sup_{L(u,v)=x} \ Min(F(u),G(v)) . \tag{4.3}$$

The condition (4.1), which is stronger than (3.6), means that L is not constant on any rectangle. It is needed only to preserve the normalization on Δ_I; thus (4.2) is essentially true for every L in $\mathcal{L}_I$. Since the operations $\tau_{Min,L}$ are associative [12], we conclude:

COROLLARY 4.2. For each L in $\mathcal{L}_I$, $\sigma_{Min,L}$ is associative on Δ_I.

Moreover, when the left end-point of I is a null element of L, $\sigma_{Min,L}$ is dual to a pointwise operation on the space of inverses of Δ_I, for then we can apply the results of [8] to get

$$[\sigma_{Min,L}(F,G)]^\wedge(t) = L(F^\wedge(t),G^\wedge(t)), \quad t \text{ in } [0,1],$$

where $F^\wedge$ is the left-continuous quasi-inverse of F, given by

$$F^\wedge(t) = \sup \ \{u: F(u) < t\}.$$

We now turn to C = Prod (and ordinal sums). It is universally known that ordinary convolution, (= $\sigma_{\text{Prod,sum}}$) is associative on Δ , *a fortiori* on Δ_+ and $\Delta_- = \Delta_{[-\infty,0]}$ since these intervals are closed under sum. And it was proved in [5] that $\sigma_{C,\text{sum}}$ is associative on Δ when C is an ordinal sum of Prod, i.e., that for L = sum the converse of Theorem 3.4 is valid. We can extend this result to more general L, in fact to any L in $\mathcal{L}_I$ that is iseomorphic to sum on one of the aforementioned intervals. The appropriate "change of variables" is the following:

LEMMA 4.3. Suppose that L in $\mathcal{L}_I$ and L_1 in $\mathcal{L}_J$ are directly iseomorphic, i.e., there is a continuous and strictly increasing function h from I onto J such that

$$L(u,v) = h^{-1}L_1(h(u),h(v)) , \qquad\qquad u,v \text{ in } I. \qquad\qquad (4.4)$$

Define the mapping $\phi : \Delta_I \rightarrow \Delta_J$ by

$$\phi(F) = F \circ h^{-1}.$$

Then ϕ is a continuous, order-preserving, one-to-one correspondence, and for each C in $\mathcal{C}$,

$$\sigma_{C,L}(F,G) = \phi^{-1}[\sigma_{C,L_1}(\phi(F),\phi(G))] . \qquad\qquad (4.5)$$

Proof. The first part of the statement is obvious; here continuity refers to the topology of weak convergence. The second part is a consequence of

$$
\begin{aligned}
\phi[\sigma_{C,L}(F,G)](x) &= \sigma_{C,L}(F,G)(h^{-1}(x)) \\
&= \iint_{L(u,v)<h^{-1}(x)} dC(F(u),G(v)) \\
&= \iint_{L_1(h(u),h(v))<x} dC(F(u),G(v)) \\
&= \iint_{L_1(y,z)<x} dC(F\circ h^{-1}(y),G\circ h^{-1}(z)) \\
&= \sigma_{C,L_1}[\phi(F),\phi(G)](x) .
\end{aligned}
$$

It is clear from (4.5) that $\sigma_{C,L}$ is associative if and only if σ_{C,L_1} is associative.

Thus, upon setting L = sum in (4.5), Lemma 4.3 and the discussion preceding it lead directly to :

THEOREM 4.4. Suppose that L in $\mathscr{L}_I$ is given by

$$L(u,v) = h^{-1}(h(u) + h(v)) , \quad u,v \text{ in } I, \tag{4.6}$$

where h is a strictly increasing function from I onto one of the intervals

$$[-\infty,\infty], \quad [0,\infty], \quad [-\infty,0].$$

Then $\sigma_{C,L}$ is associative on Δ_I if $C = $ Prod or C is an ordinal sum of Prod.

Note that any L defined via (4.6) belongs to $\mathscr{L}_I$. Associative $\sigma_{Prod,L}$ can now be constructed at will. Two key examples on Δ_+ are obtained by taking $h(u) = \log u$ and

$$h_\alpha(u) = \begin{cases} u^\alpha, & \alpha > 0, \\ -u^\alpha, & \alpha < 0. \end{cases}$$

The first gives L = mult, and the second

$$L_\alpha(u, v) = (u^\alpha + v^\alpha)^{1/\alpha}, \qquad \alpha \neq 0, \tag{4.7a}$$

whose limiting cases are

$$L_{-\infty} = \min, \quad L_\infty = \max. \tag{4.7b}$$

While by no means all L in $\mathscr{L}_I$ are of the form (4.6), (e.g., max, min), the representation theorems for associative real functions guarantee that most of the interesting L are included.

An ordinal sum of Prod is a copula C of the following form: given a set of non-overlapping subintervals $[a_i,b_i]$ of I,

$$C(s, t) = \begin{cases} a_i + (s-a_i)(t-a_i)/(b_i-a_i), & (s, t) \text{ in } [a_i,b_i]^2, \\ \text{Min}(s, t), & \text{otherwise.} \end{cases}$$

The stochastic interpretation of such copulas is left to the reader.

In the theory of probabilistic metric spaces [12], a prominent role is played by *triangle functions*: associative, commutative, order-preserving, binary operations on Δ_+

for which ϵ_0 is an identity. One can now readily obtain the triangle functions among the $\sigma_{C,L}$ by combining Theorem 4.4 and Corollary 4.2 with Theorems 2.1, 2.2, and 2.5. In particular, note that ϵ_0 is an identity for $\sigma_{Prod,L}$ if h(0) = 0 in (4.6).

Corollary 4.2 is not valid if one removes the hypothesis that L is non-decreasing; indeed,

$\sigma_{Min,mult}$ is not associative on Δ.

We present two examples, the first of which is trivial. Consider the functions $F_{-1,1}{}^s$ as given by (3.7). Then $\sigma_{Min,mult}(F_{-1,1}{}^s, F_{-1,1}{}^t) = F_{-1,1}{}^{|s+t|}$, but the function $\varphi(s,t) = |s - t|$ is not associative. For a second example, let $F = D_{2,4}$, $G = D_{-2,0}$, $H = D_{0,2}$, where $D_{u,v}$ is the uniform distribution on [u,v]. Routine calculations yield different distributions on either side of (3.1) for $\sigma_{Min,mult}$. A stochastic model is illuminating. For $-1 \le \omega \le 1$, let $X(\omega) = \omega + 3$, $Y(\omega) = \omega - 1$, $Z(\omega) = \omega + 1$. Then F,G,H are the d.f.'s of X,Y,Z, $C_{X,Y} = C_{Y,Z} = C_{X,Z} = $ Min since each is an increasing function of the others, and $C_{XY,Z} = $ Min as well, since $XY = Z^2 - 4$, $0 \le Z \le 2$. But $C_{X,YZ} \ne $ Min since $YZ = (X - 2)(X - 4)$, $0 \le X \le 4$, which is not an increasing function of X.

On the other hand the non-decreasing character of L is not necessary, for

$\sigma_{Prod,mult}$ is associative on Δ.

To verify this, for F in Δ let F^+, F^- in Δ_+ be given by

$$F^+(x) = \begin{cases} 0, & x \le 0, \\ F(x), & x > 0; \end{cases} \qquad F^-(x) = \begin{cases} 0, & x \le 0, \\ 1 - F(-x), & x > 0. \end{cases}$$

Let $*$ and $\circ$ denote the operation $\sigma_{Prod,mult}$ on Δ_+ and Δ, respectively. One can then directly establish

$$(F \circ G)^+ = F^+ * G^+ + F^- * G^-, \quad \text{and} \quad (F \circ G)^- = F^+ * G^- + F^- * G^+.$$

Now by virtue of the associativity and linearity of $*$ (Theorem 5.1), we get

$$\begin{aligned} [(F \circ G) \circ H]^+ &= (F^+ * G^+) * H^+ + (F^- * G^-) * H^+ \\ &\quad + (F^+ * G^-) * H^- + (F^- * G^+) * H^- \\ &= [F \circ (G \circ H)]^+, \end{aligned}$$

and similarly for $[(F \circ G) \circ H]^-$.

In an analogous fashion, it can be shown that σ_{Prod,L_2} is associative on Δ, where

$$L_2(u, v) = \sqrt{u^2 + v^2}.$$

Finally, the author has constructed some discontinuous (and rather uninteresting) L for which $\sigma_{C,L}$ is associative for C other than Prod or Min.

5. Further Properties; Generalized Convolutions

5.1 LINEARITY

THEOREM 5.1. Let L in $\mathscr{B}_I$ be commutative and non-decreasing in each place. The operation $\sigma_{C,L}$ is linear, i.e., for all F, G, H in Δ_I and all s, t > 0 with s + t = 1,

$$\sigma_{C,L}(sF + tG , H) = s\sigma_{C,L}(F,H) + t\sigma_{C,L}(G,H) , \qquad (5.1)$$

if and only if C = Prod.

Proof. The Lebesgue-Stieltjes integral $\iint dF(u)\, G(v)$ is linear, so $\sigma_{Prod,L}$ is linear for all L in $\mathscr{B}_I$.

To establish the converse, note first that, in view of (2.2), $\sigma_{C,min}$ is linear iff $\overline{C}$ is linear iff C is linear iff C = Prod (via Cauchy's functional equation [1]).

Now if L $\neq$ Min, there exist u,v in I with u < v such that L(u,u) < L(u,v). Set $F = \epsilon_u$, $G = \epsilon_v$, and $H = F_{uv}{}^r$ (cf. (3.7)) for r in [0,1], and note that for s + t = 1, $s\epsilon_u + t\epsilon_v = F_{uv}{}^s$. Then for all x in (L(u,u), L(u,v)], the left-hand side of (5.1) is

$$\sigma_{C,L}(F_{uv}{}^s , F_{uv}{}^r)(x) = C(s,r) ,$$

while the right-hand side is

$$s\sigma_{C,L}(\epsilon_u , F_{uv}{}^r)(x) + t\sigma_{C,L}(\epsilon_v , F_{uv}{}^r)(x) = s \cdot r + t \cdot 0 = sr.$$

Thus if $\sigma_{C,L}$ is linear, C(s,r) = sr and so C = Prod.

This result should be compared with a theorem of C. Alsina [2]: if τ is a triangle function which is linear, is continuous (in the topology of weak convergence), and for which $\tau(\epsilon_u,\epsilon_v) = \epsilon_{L(u,v)}$ for all u,v $\geq$ 0, then $\tau = \sigma_{Prod,L}$.

5.2 HOMOGENEITY

A binary operation τ on Δ_+ is said to be *homogeneous* if, for all $\lambda > 0$,

$$\tau(F_\lambda, G_\lambda) = [\tau(F, G)]_\lambda , \tag{5.2}$$

where

$$F_\lambda(x) = F(x/\lambda).$$

THEOREM 5.2. For L in $\mathcal{B}_+$, $\sigma_{C,L}$ is homogeneous if and only if L is homogeneous of degree one, i.e.,

$$L(\lambda u, \lambda v) = \lambda L(u, v), \quad \lambda > 0.$$

Moreover, if L in $\mathcal{L}_+$ satisfies the hypotheses of Theorem 4.4 with $I = [0, \infty]$, then $\sigma_{C,L}$ is homogeneous if and only if L is a member of the family (4.7a) .

Proof. The first statement follows almost immediately from (2.5), the fact that $[\epsilon_u]_\lambda = \epsilon_{\lambda u}$, and the observations

$$[\sigma_{C,L}(F, G)]_\lambda (x) = \iint_{\lambda L(u, v) < x} dC(F(u), G(v)) ,$$

$$\sigma_{C,L}(F_\lambda, G_\lambda)(x) = \iint_{L(y, z) < x} dC(F(y/\lambda), G(z/\lambda))$$
$$= \iint_{L(\lambda u, \lambda v) < x} dC(F(u), G(v)) .$$

The second part is a consequence of a famous theorem on homogeneous solutions of the associativity equation due to H.F. Bohnenblust [4]. If $h(0) = 0$ in (4.6), then 0 is an identity for L, and all of the hypotheses in [4] are satisfied, whence $L = L_\alpha$ for some $\alpha > 0$. If $h(0) = -\infty$ and $h(\infty) = 0$, let $g(u) = -h(1/u)$. Then $g(0) = 0$, and $g^{-1}(g(u)+g(v)) = 1/L(1/u, 1/v)$, which again meets all of the requirements of [4], whence $1/L(1/u, 1/v) = (u^\alpha + v^\alpha)^{1/\alpha}$, i.e., $L = L_\alpha$ for some $\alpha < 0$. The third possibility is that $h(0) = -\infty$ and $h(\infty) = \infty$. Then $h(u_0) = 0$ for some $u_0 > 0$, in which case $L(u_0, u_0) = u_0$, implying that $L(u, u) = (u/u_0)L(u_0, u_0) = u$ for all u, contradicting (4.6) .

Note that, for the limiting cases (4.7b), $\sigma_{C,max}$ and $\sigma_{C,min}$ are homogeneous. Also, σ_{C, L_α} has identity ϵ_0 or ϵ_∞, depending on whether α is positive or negative.

5.3. CONTINUITY

In Lemma 4.3, the continuity of ϕ implies that $\sigma_{C,L}$ in (4.5) is continuous if and only if σ_{C,L_1} is continuous. Now convolution ($=\sigma_{\text{Prod,sum}}$) is continuous on Δ, and Δ_+, Δ_- are closed subspaces. Identical statements are valid for $\sigma_{\text{Min,sum}}$ ($= \tau_{\text{Min,sum}}$) [12]. These facts establish:

THEOREM 5.3. Suppose that L in $\mathfrak{L}_I$ satisfies the hypotheses of Theorem 4.4. Then $\sigma_{\text{Prod,L}}$ and $\sigma_{\text{Min,L}}$ are continuous in the topology of weak convergence.

This exceedingly narrow result is the only information concerning continuity among the $\sigma_{C,L}$ which is known to the author. It is highly plausible that $\sigma_{C,L}$ is continuous for a wide class of copulas (all ?) and most continuous L. None of the usual proofs for convolution readily extend, because they ultimately rely on the linearity of Prod.

5.4 GENERALIZED CONVOLUTIONS

In a series of papers beginning in 1964 [14], K. Urbanik has extended convolution on Δ_+ in a somewhat different direction. The basic idea is to select a set of conditions just strong enough to guarantee that the resultant binary operations admit "characteristic functions", and then, within this class, to study questions of decomposability, divisibility, stability, an so on. We now compare Urbanik's system, which he calls generalized convolutions, with the $\sigma_{C,L}$.

In our notation, a *generalized convolution* is a binary operation η on Δ_+ which is commutative, associative, linear, homogeneous, and continuous, which has identity ϵ_0, and which satisfies the following "law of large numbers": there exists a sequence λ_n such that $[\epsilon_1{}^n]_{\lambda_n}$ converges weakly to F distinct from ϵ_0, where $\epsilon_1{}^n$ means $\eta(\epsilon_1,\cdots,\epsilon_1)$ (n places).

The operations $\sigma_{C,L}$ that are also generalized convolutions in this sense are few; for upon combining Theorems 3.4, 4.4, 5.1, 5.2(and the ensuing remarks), and 5.3, we have:

THEOREM 5.4. Among the operations $\sigma_{C,L}$, the only generalized convolutions are $\sigma_{\text{Prod},L_\alpha}$, $0 < \alpha \leq \infty$, where L_α is given in (4.7).

The σ_{Prod,L_α} , which Urbanik calls α-convolutions, characterize a number of properties among the generalized convolutions -- see [10] and the references cited therein. A consequence of Theorem 5.4 is yet another such characterization: the α-convolutions are the only generalized convolutions that are induced by operations on random variables.

5.5 INEQUALITIES

Inequalities among the $\sigma_{C,L}$ and between the $\sigma_{C,L}$ and other families of operations on Δ have been studied extensively [12]. In particular, sharp bounds for $\sigma_{C,sum}$, which are useful in calculations, are developed in [7]. (See Schweizer's contribution to this volume.)

5.6 SOME OPEN QUESTIONS

Perhaps the most basic unsolved structural problem is to determine precisely which $\sigma_{C,L}$ are continuous, as discussed in Section 5.3, or at least to establish continuity for key families of copulas.

A second problem, having stochastic implications, deals with the question of uniqueness, in the following sense:

$$\sigma_{C_1,L}(F,G) = H = \sigma_{C_2,L}(F,G) \quad \rightarrow \quad C_1 = C_2 .$$

The exact meaning here has intentionally been left vague -- in particular the quantification over F,G,H, and L. The problem has been solved in [3] for the special case where L = sum and F,G,H are uniform. In this regard, we offer a relevant example. Let X and Z be uniform on [-1,1], but otherwise arbitrary, let Y = -X, and take $L(u,v) = |u| + |v|$. Clearly, $L(X,Z) = L(Y,Z)$, but $C_{Y,Z}(s,t) = t - C_{X,Z}(1-s,t)$.

A third problem -- a very general one -- is to explore arithmetic properties in the family $\sigma_{C,L}$ (idempotent elements, decomposability, ...).

References

[1]   Aczél, J. and Dhombres, J. (1989) Functional Equations in Several Variables, Cambridge University Press, Cambridge.

[2]   Alsina, C. (1985) 'A characterization of convolution and related

operations', Aequationes Math. 28, 88-93.

[3]   Alsina, C. and Bonet, E. (1979) 'On sums of dependent uniformly distributed random variables', Stochastica 3, 33-43.

[4]   Bohnenblust, H.F. (1940) 'An axiomatic characterization of L_p-spaces', Duke Math. J. 6, 627-640

[5]   Frank, M.J. (1975) 'Associativity in a class of operations on spaces of distribution functions', Aequationes Math. 12, 121-144.

[6]   Frank, M.J. (1979) 'On the simultaneous associativity of F(x,y) and x + y - F(x,y)', Aequationes Math. 19, 194-226.

[7]   Frank, M.J., Nelsen, R.B., and Schweizer, B. (1987) 'Best-possible bounds for the distribution of a sum - a problem of Kolmogorov', Probab. Th. Rel. Fields 74, 199-211.

[8]   Frank, M.J. and Schweizer, B. (1979) 'On the duality of generalized infimal and supremal convolutions', Rend. Math. (6) 12, 1-23.

[9]   Genest, C. (1987) 'Frank's family of bivariate distributions', Biometrika 74, 549-555.

[10]   Kucharczak, J. (1988) 'Decomposability of point measures in generalized convolution algebras', Colloq. Math. 55, 163-167.

[11]   Nelsen, R.B. (1986) 'Properties of a one-parameter family of bivariate distributions with specified marginals', Commun. Statist. - Theory Meth. 15, 3277-3285.

[12]   Schweizer, B. And Sklar, A. (1983) Probabilistic Metric Spaces, Elsevier North-Holland, New York.

[13]   Sklar, A. (1959) 'Fonctions de répartition à n dimensions et leur marges', Publ. Inst. Statist. Univ. Paris 8, 229-231.

[14]   Urbanik, K. (1964) 'Generalized convolutions', Studia Math. 23, 217-245.

PROBABILISTIC INTERPRETATIONS OF COPULAS AND THEIR CONVEX SUMS

P. MIKUSIŃSKI, H. SHERWOOD, and M.D. TAYLOR
Department of Mathematics, University of Central Florida
Orlando, FL 32816-6990
U.S.A.

This work was partially supported by a grant from the University of Central Florida Division of Sponsored Research In-House Grants Program.

ABSTRACT. The copula C contains valuable information about the type of dependence that exists between random variables having C as their copula. It is this dependence, captured in the copula, which is of interest in this paper. One might think of a copula as a canonical representative of all those distribution functions H that correspond to random variables X and Y which have a specific sort of relationship to each other. It would be nice to have a catalog of copulas which specifies the sort of dependence each copula captures. With such a long term goal in mind, we give probabilistic interpretations for certain copulas, i.e., we give necessary and sufficient conditions on the random variables X and Y in order that C should be their copula. Also we give probabilistic interpretations for convex sums (mixtures) of copulas.

1. Introduction

Throughout this paper X and Y denote random variables defined on the same probability space, and F, G, and H denote the distribution functions of X, Y and (X,Y), respectively. Whenever X and Y bear affixes, F, G, and H bear those same affixes. Also, I, $\mathbb{R}$ and $\bar{\mathbb{R}}$ denote the closed unit interval [0,1], the reals and the extended reals, respectively. The word "rectangle" refers to the Cartesian product of two intervals either or both of which may be degenerate, i.e., a single point. We shall let μ_H denote the Lebesgue-Stieltjes measure induced by H. Finally, to minimize technicalities, we assume in the first four sections of this paper that both F and G are continuous.

In 1959 A. Sklar [12] introduced the notion of a *copula* which is a function $C: I^2 \to I$ such that

$$C(s,0) = 0 = C(0,s) \text{ and } C(s,1) = s = C(1,s) \tag{1.1}$$
whenever $0 \le s \le 1$, and

$$C(s_2,t_2) - C(s_2,t_1) + C(s_1,t_1) - C(s_1,t_2) \geq 0 \qquad (1.2)$$

whenever $0 \leq s_1 \leq s_2 \leq 1$ and $0 \leq t_1 \leq t_2 \leq 1$.

It is helpful to think of a copula in this way: A unit mass is spread across the unit square subject to the conditions that the amount of mass to the left of any vertical line $x = a$ (where $0 \leq a \leq 1$) is a and the amount of mass below any horizontal line $y = b$ (where $0 \leq b \leq 1$) is b. Then $C(a,b)$ is the amount of mass in the rectangle $[0, a] \times [0, b]$. The Lebesgue-Stieltjes measure μ_C induced by C is a *doubly stochastic measure* because $\mu_C(A \times I) = \mu_C(I \times A) = \lambda(A)$ for each Borel subset A of I where λ denotes Lebesgue measure.

Also in [12], Sklar presents the following result: Corresponding to X and Y there is a unique copula C, called the *copula* for (X, Y), such that

$$H(x,y) = C(F(x), G(y)) \text{ for all } x, y \text{ in } \mathbb{R}. \qquad (1.3)$$

(If F or G fails to be continuous, the word "unique" must be deleted in the preceding statement.)

The copula C contains valuable information about the type of dependence that exists between random variables having C as their copula. It is this dependence, captured in the copula, which is of interest in this paper. One might think of a copula as a canonical representative of all distribution functions H that correspond to random variables X and Y which have a specific sort of relationship to each other. It would be nice to have a catalog of copulas which specifies the sort of dependence each copula captures. With such a long term goal in mind, we give probabilistic interpretations for certain copulas, i.e., we give necessary and sufficient conditions on the random variables X and Y in order that C should be their copula. Also we give probabilistic interpretations for convex sums (mixtures) of copulas.

2. The Most Important Copulas

The most important copulas are Π, M and W, defined on I^2 as follows:

$$\Pi(s,t) = s \cdot t, \qquad (2.1)$$

$$M(s,t) = \text{Min}(s,t), \qquad (2.2)$$

$$W(s,t) = \text{Max}(s + t - 1, 0). \qquad (2.3)$$

Of these three, Π is most important because X and Y are independent if and only if $H(x,y) \equiv F(x)G(y)$. Thus we have the following:

***Probabilistic Interpretation of* Π.** Π is the copula for (X, Y) if and only if X and Y are independent.

In 1951 M. Fréchet [3] established what have come to be known as the **Fréchet bounds**, namely that for any real numbers x and y,

$$W(F(x), G(y)) \leq H(x,y) \leq M(F(x), G(y)). \qquad (2.4)$$

A natural question arises—under what conditions do the inequalities in (2.4) become identities? In the same paper Fréchet answered this question thereby giving probabilistic interpretations for M and W. We shall not only give these interpretations, but we shall provide some heuristic motivation for them. It should be mentioned that a number of authors, including G. Dall'Aglio [2], G. Kimeldorf and A.R. Sampson [5], and E. F. Wolff [14] have given proofs (some more detailed than others) of these probabilistic interpretations for M and W. We like the approach given below, which is a modification of Wolff's approach, because it is so geometric and can be used in other situations.

Suppose C is the copula for (X, Y) and consider the probability transform $\phi: \overline{\mathbb{R}}^2 \to I^2$, defined as follows:

$$\phi(x, y) = (F(x), G(y)). \tag{2.5}$$

Observe that ϕ maps $\overline{\mathbb{R}}^2$ onto I^2 in such a way that images (and preimages) of rectangles are rectangles, right and left are preserved as are above and below. Moreover, ϕ is measure preserving with respect to the measures μ_C and μ_H.

Now suppose M is the copula for (X, Y). It is easy to see that M has its unit mass spread uniformly along D, the diagonal from $(0,0)$ to $(1,1)$ shown in Figure 1.

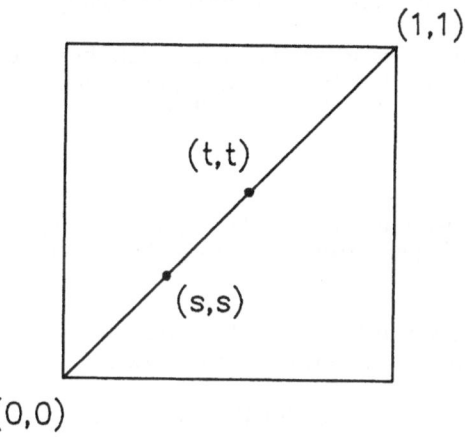

Figure 1. Illustration of diagonal D where mass of M is concentrated.

Since the probability transform ϕ is measure preserving, the mass of the distribution H will be concentrated on the preimage of D. To see what $\phi^{-1}(D)$ looks like we first recall that the preimage of the rectangle $\{(s,s)\}$ will again be a rectangle (which may also be degenerate, i.e., a singleton point, a vertical line segment, or a horizontal line segment). In fact for all but a countable number of values of s, the preimage of $\{(s,s)\}$ will again be a singleton point. If $0 \le s < t \le 1$, then the fact that (s,s) is strictly below and strictly to the left of (t,t) implies that the preimage of $\{(s,s)\}$ is strictly below and strictly to the left of the preimage of $\{(t,t)\}$. Thus the preimage of D under ϕ will look something like the picture in Figure 2.

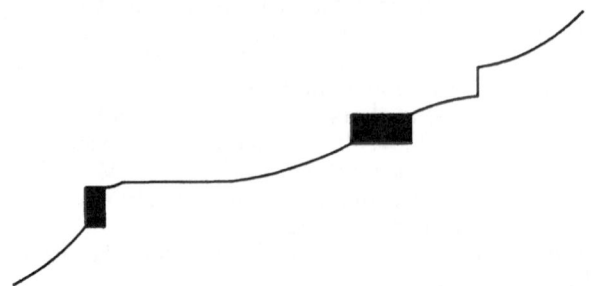

Figure 2. Illustration of $\phi^{-1}(D)$ where mass of H is concentrated.

Whenever the preimage of $\{(s,s)\}$ is a vertical (horizontal) line segment we remove from $\phi^{-1}(D)$ all but the bottom (left) point of that segment. Whenever the preimage of $\{(s,s)\}$ is a nondegenerate rectangle we remove all but the lower, left corner. Since each set removed has μ_H-measure zero and there are at most a countable number of such sets, the μ_H-measure of what remains is still one. What remains of $\phi^{-1}(D)$ is the graph of a strictly increasing function f and we see that $Y = f(X)$, almost surely.

In the other direction, suppose that f is a strictly increasing function such that $Y = f(X)$, almost surely. Let (x,y) be any point such that $y = f(x)$. Further, let $F(x) = r$. Observe that $G(y) = r$ as well. Then,

$$\phi(x,y) = (F(x), G(y)) = (r,r).$$

It now follows that the mass which the copula for (X, Y) deposits in I^2 is spread uniformly along the diagonal from (0,0) to (1,1). In other words, M is the copula for (X, Y).

***Probabilistic Interpretation of* M.** M is the copula for (X, Y) if and only if X and Y are almost surely increasing functions of each other.

Similar motivation can be given for the following:

***Probabilistic Interpretation of* W.** W is the copula for (X, Y) if and only if X and Y are almost surely decreasing functions of each other.

3. Shuffles of Min

Instead of giving an analytic definition for a shuffle of Min, we shall give a prescription for constructing its mass distribution. First, start with the mass distribution for M which, recall, has its unit mass spread uniformly on that diagonal of I^2 having positive slope. Second, cut I^2 vertically into a finite number of strips. Third, shuffle the strips with perhaps some of them flipped around their vertical axes of symmetry. Fourth, reassemble the strips to reform the unit square. This construction produces the mass distribution of a copula called a *shuffle of Min*. An example is pictured in Figure 3.

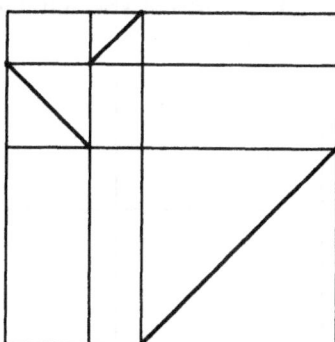

Figure 3. Illustration of set where mass is concentrated for a shuffle of Min.

In [5] G. Kimeldorf and A. K. Sampson prove essentially that Π, the copula for independence, can be approximated arbitrarily closely by certain shuffles of Min. The authors of this paper have proved that any copula can be uniformly approximated arbitrarily closely by certain shuffles of Min. This means that the behavior of any two continuous random variables can be approximated so closely by two that are invertible functions of each other that it would be impossible, experimentally, to distinguish one pair from the other. We exhibit a very short and elegant proof of this result below. R. A. Vitale independently proved a similar result in [13].

In [6] A. W. Marshall addresses the following question: Which bivariate distributions with uniform marginals are important in their own right? His approach to this question leads to a large class of bivariate uniform distributions which are characterized in terms of certain shuffles of Min. More will be said about this later.

To see what the probabilistic interpretation of a shuffle of Min should be, suppose C is the shuffle of Min whose support set is pictured in Figure 3. Suppose further that C is the copula for (X, Y). Each vertical strip is a rectangle whose preimage under the probability transform ϕ, defined in (2.5), will be a vertical strip running from the bottom to the top of $\overline{\mathbb{R}}^2$. Moreover the order of the strips from left to right is preserved. Similar observations can be made concerning the horizontal strips shown in Figure 3. Observe that in each vertical and each horizontal strip there is one and only one rectangle having a diagonal with positive mass spread on it. If D is such a diagonal and D has positive slope, then $\phi^{-1}(D)$ less a set of μ_H-measure zero will be the graph of an increasing function. If, on the other hand, the diagonal D has negative slope, then $\phi^{-1}(D)$ less a set of μ_H-measure zero will be the graph of a decreasing function. Each of the vertical strips and each of the horizontal strips in $\overline{\mathbb{R}}^2$ contain exactly one preimage of a diagonal with positive mass on it (see Figure 4).

The union of the preimages of these diagonals less a set of μ_H-measure zero is the graph of a function f. Moreover, Y = f(X), almost surely.

This is a good picture of what happens in general. We shall call a function having the sort of graph displayed in Figure 4 a *strongly piecewise strictly monotone function*. Obviously such a function is piecewise strictly monotone but we use the modifier "strongly" to distinguish this sort of function from one like $g(x) = x^{-2}$ which is piecewise strictly monotone but not strongly piecewise strictly monotone.

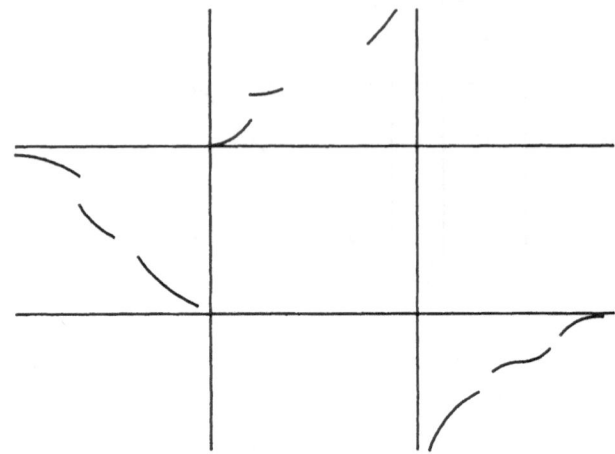

Figure 4. Illustration of mass concentration of H when copula for H is a shuffle of Min.'

More precisely, we say that a function f is *strongly piecewise strictly monotone* if and only if $\overline{\mathbb{R}}^2$ can be partitioned into a finite number of rectangles such that in each column and in each row of rectangles there is exactly one rectangle having a nonempty intersection with the graph of f and that portion of the graph of f is strictly monotone.

In the opposite direction it can be shown that the copula for (X, Y) is a shuffle of Min whenever X and Y are strongly piecewise strictly monotone functions of each other. We may now state the following:

Probabilistic Interpretation of a Shuffle of Min. The copula for (X, Y) is a shuffle of Min if and only if X and Y are strongly piecewise strictly monotone functions of each other.

Theorem 3.1 The shuffles of Min are dense in the set of all copulas endowed with the sup norm.

Proof: Let C be an arbitrary copula and let $\epsilon > 0$ be given. We shall construct C', a shuffle of Min, such that $\| C - C' \| < \epsilon$. Using Lemma 6.1.9 of [10] we may choose a positive integer K such that for any copula C^*,

$$| C^*(s,t) - C^*(u,v) | < \tfrac{\epsilon}{2} \quad \text{whenever} \quad |s - u| < \tfrac{1}{K} \text{ and } |t - v| < \tfrac{1}{K}. \quad (3.1)$$

Next, subdivide I^2 into K vertical columns and K horizontal rows as follows:

$$V_i = [(i-1)/K, i/K] \times I \quad \text{and} \quad H_j = I \times [(j-1)/K, j/K]$$

for $i, j = 1, 2, \ldots, K$. Set $S_{ij} = V_i \cap H_j$ and let $m_{ij} = \mu_C(S_{ij})$. Since μ_C is doubly stochastic,

$$\mu_C(V_i) = m_{i1} + m_{i2} + \cdots + m_{iK} = 1/K$$

and

$$\mu_C(H_j) = m_{1j} + m_{2j} + \cdots + m_{Kj} = 1/K.$$

Subdivide each V_i into K vertical subcolumns, labeled from left to right V_{i1}, $V_{i2}, \ldots, V_{iK}$, so that the width of each V_{ik} is m_{ik}. Similarly subdivide each H_j into K horizontal subrows, labeled from bottom to top $H_{j1}, H_{j2}, \ldots, H_{jK}$, so that the height of each H_{jk} is m_{kj}. Then $V_{ij} \cap H_{ji}$ is a square with sides of length m_{ij} located in S_{ij}. For each $i, j = 1, 2, \ldots, K$, spread a mass of size m_{ij} uniformly along the diagonal of $V_{ij} \cap H_{jj}$ which has positive slope. This is clearly a mass distribution for a shuffle of Min which we denote by C'. Since $\mu_{C'}(S_{ij}) = m_{ij} = \mu_C(S_{ij})$ we have $C(i/K, j/K) = C'(i/K, j/K)$ for $i, j = 0, 1, \ldots, K$. Finally, let $(x, y) \in I^2$. There is some $i, j = 0, 1, \ldots, K$ such that $|x - i/K| < 1/K$ and $|y - j/K| < 1/K$. Thus, by (3.1) we have

$$| C(x,y) - C'(x,y) | \leq | C(x,y) - C(i/K, j/K) |$$

$$+ | C(i/K, j/K) - C'(i/K, j/K) |$$

$$+ | C'(i/K, j/K) - C'(x,y) |$$

$$< \tfrac{\epsilon}{2} + 0 + \tfrac{\epsilon}{2} = \epsilon,$$

and the proof is complete.

The following result is an immediate consequence of the preceding one.

<u>Theorem 3.2</u>   Given any $\epsilon > 0$ and any X, Y, there exist X^* and Y^* and an invertible piecewise increasing function f such that $Y^* = f(X^*)$, $F = F^*$, $G = G^*$, and $\| H - H^* \| < \epsilon$ where $\| \cdot \|$ denotes the sup-norm.

As mentioned earlier, R. A. Vitale [13] has proved a different version of this result and has generalized it to more than two random variables. His version is especially nice in that it lends itself nicely for simulation purposes. The reader should look elsewhere in these proceedings for additional work in this vein by Vitale.

4. Generalized Hairpins

The set of all copulas is a convex set, i.e., whenever C_1 and C_2 are copulas and p_1 and p_2 are nonnegative numbers with $p_1 + p_2 = 1$ we have $p_1 C_1 + p_2 C_2$ is also a copula. A copula C is *extremal* if it has the property that p_1 or p_2 is zero whenever $C = p_1 C_1 + p_2 C_2$ and $C_1 \neq C_2$. The extremal ones are of special interest. In [11] T. L. Seethoff and R. C. Shiflett and in [4] A. Kamiński and two of the authors of this paper essentially show that at most one copula can have its mass concentrated on the union of the graphs of two increasing functions and hence that copula must be extremal.

A *generalized hairpin* is the union of the graphs of $y = L(x)$ and $x = U(y)$ where L and U are nondecreasing functions from I into I such that

$$(L \circ U)(t) < t \text{ and } (U \circ L)(t) < t \text{ whenever } 0 < t < 1. \tag{4.1}$$

The geometric meaning of condition (4.1) is that the graph of $x = U(y)$ is above and to the left of the graph of $y = L(x)$. (See Figure 5 for a sketch.)

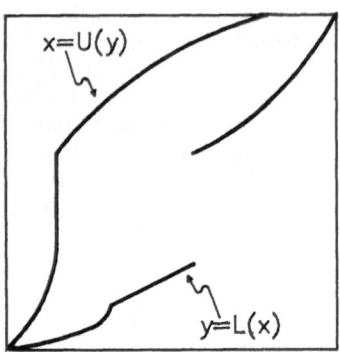

Figure 5. Illustration of generalized hairpin.

We now present a probabilistic interpretation for copulas having mass concentrated on generalized hairpins.

Probabilistic Interpretation of a Copula with Mass on a Generalized Hairpin. The copula for (X, Y) has its mass concentrated on a generalized hairpin if and only if the disjunction, $Y = f(X)$ or $X = g(Y)$, is almost surely true for some nondecreasing functions $f, g: \overline{\mathbb{R}} \to \overline{\mathbb{R}}$ satisfying

$$(f \circ g)(y) < y \text{ whenever } 0 < G(y) < 1 \tag{4.2}$$

and

$$(g \circ f)(x) < x \text{ whenever } 0 < F(x) < 1. \tag{4.3}$$

Proof: Suppose the copula C for (X, Y) has its mass concentrated on a generalized hairpin which is the union of the graphs of $y = L(x)$ and $x = U(y)$ where L and U are nondecreasing functions from I into I satisfying (4.1). Let ϕ be the probability transform defined in (2.5). The graphs of $y = f(x)$ and $x = g(y)$ will turn out to be essentially the preimages under ϕ of the graphs of $y = L(x)$ and $x = U(y)$. Let (u,v) be any arbitrary point on the graph of L. Then the set $\phi^{-1}(\{(u,v)\})$ is a closed, possibly degenerate, rectangle. Notice that the rectangle $\phi^{-1}(\{(u,v)\})$ must be finite in its extent if (u,v) is an interior point of I^2. This follows from the definition of ϕ and the fact that μ_C must have a positive amount of mass to either side of (u,v) and above and below it. We construct f as follows:

(a)   If $\phi^{-1}(\{(u,v)\}) = \{(x,y)\}$, then (x,y) is on the graph of f.

(b)   If $\phi^{-1}(\{(u,v)\})$ is a vertical line segment, then the bottom point (x,y) is on the graph of f.

(c)   If $\phi^{-1}(\{(u,v)\})$ is a horizontal line segment, then it is all part of the graph of f.

(d)   If $\phi^{-1}(\{(u,v)\})$ is a nondegenerate rectangle, then the line segment from the lower left corner to the upper right corner is contained in the graph of f.

(In cases when the rectangle is infinite in extent, which can only happen when $(u,v) = (0,0)$ or $(1,1)$, then either the upper right or the lower left corner must be finite. We then take our "line segment" to be the half-infinite ray which proceeds across the rectangle from this corner at a 45° angle from the horizontal. Of course we set $f(-\infty) = -\infty$ and $f(\infty) = \infty$.)

The continuity of F ensures that the domain of f is all of $\bar{\mathbb{R}}$. Clearly f is non-decreasing. Let us identify f and L with their graphs. Then, $\mu_H(\phi^{-1}(L)) = \mu_C(L)$, the graph of f differs from $\phi^{-1}(L)$ at most on the nonsingleton $\phi^{-1}(\{(u,v)\})$, there exist only a countable number of these last objects and each of them has μ_H-measure zero. Therefore we must have $\mu_H(f) = \mu_C(L)$.

We construct g in a similar way by considering the preimage under ϕ of the graph of $x = U(y)$ and as before obtain $\mu_H(g) = \mu_C(U)$ where we have identified g and U with the graphs of $x = g(y)$ and $x = U(y)$. Thus,

$$P[Y = f(X) \text{ or } X = g(Y)] = \mu_H(L) + \mu_H(U) = 1.$$

Note that when considering points on the graphs of $y = L(x)$ and $x = U(y)$, if (u,v) is above (or to the left of) (r,s), then $\phi^{-1}(\{(u,v)\})$ is above (or to the left of) $\phi^{-1}(\{(r,s)\})$. This implies that $(f \circ g)(y) < y$ whenever $0 < G(y) < 1$, and $(g \circ f)(x) < x$ whenever $0 < F(x) < 1$. This gives the result in one direction.

Now suppose there exist nondecreasing functions $f, g : \bar{\mathbb{R}} \rightarrow \bar{\mathbb{R}}$ such that the disjunction, $Y = f(X)$ or $X = g(Y)$, is almost surely true and (4.2) and (4.3) are satisfied. Identify f and g with the graphs of $y = f(x)$ and $x = g(y)$. Let us start by defining $L^* = \phi(f)$. By the continuity of F we have $\{u : (u,v) \in L^*\} = I$, the entire unit interval. To show that L^* is "nondecreasing", choose (u_1, v_1) and (u_2, v_2) from L^* such that $u_1 < u_2$. It must be possible to find (x_1, y_1) and (x_2, y_2) in f such that $\phi(x_i, y_i) = (u_i, v_i)$ and $x_1 < x_2$. Since f is nondecreasing we see that $y_1 \leq y_2$ and so $v_1 \leq v_2$.

The set L^* may fail to be a function. To see why, observe that as x increases, $(x, f(x))$ traces out the graph of f and $(F(x), G(f(x)))$ traces out L^*. Over some interval where F is constant, $G \circ f$ may be increasing so that $(F(x), G(f(x)))$ will climb vertically. The graph of F can have at most a countable number of these flat spots so L^* can have at most a countable number of these vertical segments. To extract the graph of the desired function L we delete from L^* all but the bottom point of each of these vertical segments. Technically we accomplish this by defining L as follows: for each u in I

$$L(u) = G(f(x_0)) \tag{4.4}$$

where x_0 is the least element of $\bar{\mathbb{R}}$ such that $F(x_0) = u$. Notice that

$$(L \circ F)(x) = (G \circ f)(x_0) \tag{4.5}$$

whenever x_0 is the least element of $\bar{\mathbb{R}}$ such that $F(x_0) = F(x)$. Then L is a subset of

L^* (we are using the definition of ϕ and identifying L with its graph), the domain of L is all of I (by the continuity of F), and L is nondecreasing (since L^* is "nondecreasing"). We define U in a similar way; it is required to satisfy

$$(U \circ G)(y) = (F \circ g)(y_0) \tag{4.6}$$

where y_0 is the least element of $\overline{\mathbb{R}}$ satisfying $G(y_0) = G(y)$.

We now show that $U \circ L$ satisfies the desired inequality presented in (4.1). To this end, suppose $0 < u < 1$. Let x_0 be the least element of $\overline{\mathbb{R}}$ such that $F(x_0) = u$ and let y_0 be the least element of $\overline{\mathbb{R}}$ such that $G(y_0) = G(f(x_0))$. Notice that $0 < F(x_0) < 1$. Consequently,

$$(U \circ L)(u) = (U \circ G \circ f)(x_0) \qquad \text{[see (4.4)]}$$

$$= (U \circ G)(y_0) \qquad \text{[see the definition of } y_0]$$

$$= (F \circ g)(y_0) \qquad \text{[see (4.6)]}$$

$$\leq (F \circ g \circ f)(x_0). \qquad \text{[see the definition of } y_0]$$

We know that $(g \circ f)(x_0) < x_0$ because $0 < F(x_0) < 1$. This implies, using the definition of x_0, that $(F \circ g \circ f)(x_0) < F(x_0) = u$. Therefore $(U \circ L)(u) < u$. In the same way, $(L \circ U)(v) < v$ whenever $0 < v < 1$. Let C denote the copula for (X, Y). Since ϕ is measure preserving, $\mu_C(L^*) = \mu_H(f)$. Because L^* differs from L by at most a countable number of line segments and each of these has μ_C-measure zero, we must have $\mu_C(L^*) = \mu_C(L)$. Thus $\mu_C(L) = \mu_H(f)$ and, in like manner, $\mu_C(U) = \mu_H(g)$. Finally, we observe that $\mu_C(L \cup U) = \mu_H(f) + \mu_H(g) = 1$. Thus the mass of C is concentrated on a generalized hairpin.

The following example presents a physical situation which illustrates the preceding result.

Example Imagine we have a 2-way channel or link running between points A and B and at A we are able to read some nonnegative numerical quantity X and at B some nonnegative numerical quantity Y. (See Figure 6)

Figure 6. Illustration of a 2-way channel running between points A and B.

The quantities X and Y might, for example, be voltage or water pressure. We assume we may specify either X or Y but not both simultaneously. If we specify X, then we will think of Y as being X decreased by some kind of "friction" or "resistance" after propagation along the link between A and B. If, on the other hand, it is Y that is specified, then it is X which is a decreased or "degraded" copy of Y. If X and Y are random variables, then their copula will have its mass concentrated on a generalized hairpin.

R. B. Nelsen [9] has calculated nonparametric measures of correlation for any pair (X,Y) of random variables whose copula has its mass on certain generalized hairpins.

5. Convex Sums of Copulas

In this section we drop the requirement that the random variables X and Y be continuously distributed.

A convex sum of copulas is easily seen to be a copula again and it is our purpose in this section to investigate the probabilistic interpretation of such convex sums. In the finite case one begins with copulas C_1, $C_2, \ldots$, C_n and positive numbers p_1, $p_2, \ldots$, p_n where $p_1 + p_2 + \cdots + p_n = 1$, and forms a new copula C via

$$C = p_1 C_1 + p_2 C_2 + \cdots + p_n C_n.$$

More generally one begins with an indexed family of copulas $\{C_z \colon z \in \mathbb{R}\}$ and defines a copula C via

$$C = \int_{\mathbb{R}} C_z(x,y) \, dP_1(z).$$

where P_1 is a probability measure on $\mathbb{R}$ (or some other probability space).

First we shall consider an example which will illustrate some of the ideas to come concerning convex sums (or mixtures) of copulas. Suppose an urn contains two balls numbered 1 and 2 respectively and a second urn contains two balls numbered 3 and 4 respectively. We also have a coin which yields heads with probability p and tails with probability $q = 1 - p$. Toss the coin and draw a ball from each urn. Let X be the number on the ball drawn from the first urn, so $X = 1$ or 2. If the coin is heads, let $Y = X + 2$; so $Y = 3$ or 4 and is an increasing function of X. If, on the other hand, the coin is tails, let Y denote the number on the ball drawn from the second urn; so $Y = 3$ or 4 and is independent of X. It is easy to verify that

$$P[X < x \text{ and } Y < y \mid \text{heads}] = M(F(x), G(y))$$

and

$$P[X < x \text{ and } Y < y \mid \text{tails}] = F(x) G(y)$$

and that the copula $C = pM + q\Pi$ is a copula for X and Y.

In general, to say that $p_1 C_1 + p_2 C_2 + \cdots + p_n C_n$, where the C_i's are copulas and the p_i's are positive numbers such that $p_1 + p_2 + \cdots + p_n = 1$, is a copula for two random variables X and Y is to say there ought to be another random variable Z, an analog to the coin of our example, with the property that Z can take on n different values, the ith of the values occurring with probability p_i, and that for all experiments performed when Z has assumed this i^{th} value, the joint distribution function for X and Y behaves as though C_i is a connecting copula. But a difficulty arises here. Such a Z may fail to exist; the probability space on which X and Y are defined may somehow be too "small" to support the desired Z. To see this, consider the probability space in which

$$\Omega = \{(1,3), (1,4), (2,3), (2,4)\},$$

the σ-algebra consists of all subsets of Ω, and the probability measure P assigns to each simple event in Ω the value given in Table 1.

Table 1. Definition of P

ω	$P(\omega)$
(1,3)	3/8
(1,4)	1/8
(2,3)	1/8
(2,4)	3/8

Define X and Y on Ω by $X(x,y) = x$ and $Y(x,y) = y$. It is easy to verify that

$$P[X = 1] = P[X = 2] = P[Y = 3] = P[Y = 4] = \tfrac{1}{2}$$

and that

$$C = \tfrac{1}{2}M + \tfrac{1}{2}\Pi$$

is a copula for (X, Y). But there is no room in this example for a coin, a random variable taking on two values, $Z: \Omega \to \{0,1\}$, which separates the two states of dependency between X and Y so that

$$P[X < x \text{ and } Y < y \mid Z = 0] = M(F(x), G(y))$$

and

$$P[X < x \text{ and } Y < y \mid Z = 1] = F(x)\,G(y).$$

If there were, then we would have

$$P[(X,Y) = (1,3) \mid Z = 0] = \tfrac{1}{2}$$

and

$$P[(X,Y) = (1,3) \mid Z = 1] = \tfrac{1}{4};$$

but this amounts to trying to split the simple event (1,3) into two disjoint events of positive probability, an impossibility.

It is possible to circumvent this situation in two different ways. One can completely change the probability space under consideration or one can "enlarge" the probability space by considering it as one factor of a product space. In both of these ways enough room is created to define an appropriate Z. In this paper we choose to do the former because it is less technical.

In our treatment of convex sums of copulas, we begin with the case of finite sums. In this case it is useful to consider random variables which are mixtures of independent random variables in the following sense:

$$X = \sum_{i=1}^{n} \delta_i(Z)\,X_i$$

where $X_1, X_2, \ldots, X_n$, and Z are independent random variables and at any given point in the probability space exactly one of the $\delta_i(Z)$'s is 1 and all the others are

0. In other words, X behaves as though part of the time it is X_1, part of the time it is X_2, etc. ... , and the random variable Z singles out the times for these different behaviors. This is a weighted average of the X_i's with the weights being the probabilities that $\delta_i(Z) = 1$ for $i = 1, 2, \ldots, n$.

In our first result on convex sums of copulas, n denotes an arbitrary but fixed natural number, each C_i is a copula, each p_i is positive and the sum of the p_i's is 1. Also, for each $i = 1, 2, \ldots, n$, we let δ_i be the function defined on $\mathbb{R}$ via

$$\delta_i(x) = \begin{cases} 1, & \text{if } x = i, \\ 0, & \text{otherwise.} \end{cases}$$

In the second half of the proof we switch from an original probability space to a new probability space of the form $\mathbb{R}^{2n} \times \{1, 2, \ldots, n\}$ in order to have "enough room" to find not only Z but also $X_1, X_2, \ldots, X_n, Y_1, Y_2, \ldots, Y_n$.

Probabilistic Interpretation of a Finite Convex Sum of Copulas:

$\sum\limits_{i=1}^{n} p_i C_i$ is a copula for (X, Y)

if and only if there exist random variables $X_1, X_2, \ldots, X_n, Y_1, Y_2, \ldots, Y_n, Z$ such that

(1) $(X_1, Y_1), (X_2, Y_2), \ldots, (X_n, Y_n)$, and Z are independent,

(2)   The X_i's are identically distributed as are the Y_i's,

(3) $P[Z = i] = p_i$ for each i,

(4)   Each C_i is a copula for (X_i, Y_i),

(5) $H = H^*$ where $X^* = \sum\limits_{i=1}^{n} \delta_i(Z) X_i$ and $Y^* = \sum\limits_{i=1}^{n} \delta_i(Z) Y_i$.

(Remember H^* is the distribution function for (X^*, Y^*).)

<u>Proof</u>:   Suppose random variables $X_1, X_2, \ldots, X_n, Y_1, Y_2, \ldots, Y_n, Z$ satisfy (1)– (5). Notice first that

$$F(x) = F^*(x) = P[X^* < x] = \sum_{j=1}^{n} P[X^* < x, Z = j] = \sum_{j=1}^{n} p_j P[X_j < x]$$

$$= \sum_{j=1}^{n} p_j F_j(x) = F_1(x) = F_2(x) = \cdots = F_n(x).$$

Similarly, $G(y) = G^*(y) = G_1(y) = G_2(y) = \cdots = G_n(y)$. Finally,

$$H(x,y) = H^*(x,y) = P[X^* < x, Y^* < y] = \sum_{j=1}^{n} P[X^* < x, Y^* < y, Z = j]$$

$$= \sum_{j=1}^{n} P[X_j < x, Y_j < y, Z = j] = \sum_{j=1}^{n} p_j H_j(x,y)$$

$$= \sum_{j=1}^{n} p_j\, C_j(F_j(x), G_j(y)) = \sum_{j=1}^{n} p_j\, C_j(F(x), G(y)).$$

To prove the reverse implication assume that $\sum_{i=1}^{n} p_i\, C_i$ is a copula for (X, Y). For each $i = 1, 2, \ldots, n$, we define K_i on $\mathbb{R}^2$ via

$$K_i(x,y) = C_i(F(x), G(y)).$$

Then K_i is a 2-dimensional distribution function. Let $(\Omega_i, \mathcal{A}_i, \mu_i)$ denote the probability space obtained by letting $\Omega_i = \mathbb{R}^2$, letting $\mathcal{A}_i$ be the σ-algebra of Borel-measurable subsets of $\mathbb{R}^2$, and letting μ_i be the Lebesgue-Stieltjes measure induced on $\mathbb{R}^2$ by K_i. Let $\Omega_{n+1} = \{1, 2, \ldots, n\}$, let $\mathcal{A}_{n+1}$ be the set of all subsets of Ω_{n+1}, and let μ_{n+1} be the measure on Ω_{n+1} for which $\mu_i(\{i\}) = p_i$ for $i = 1, 2, \ldots, n$. Let $(\Omega, \mathcal{A}, P)$ be the product of the probability spaces $(\Omega_i, \mathcal{A}_i, \mu_i)$ for $i = 1, 2, \ldots, n+1$. Next we define random variables $X_1, X_2, \ldots, X_n, Y_1, Y_2, \ldots, Y_n$ and Z on Ω as follows: For $i = 1, 2, \ldots, n$,

$$X_i((x_1,y_1), (x_2,y_2), \ldots, (x_n,y_n), z) = x_i,$$

$$Y_i((x_1,y_1), (x_2,y_2), \ldots, (x_n,y_n), z) = y_i,$$

and

$$Z((x_1,y_1), (x_2,y_2), \ldots, (x_n,y_n), z) = z.$$

It is immediate that $(X_1, Y_1), (X_2, Y_2), \ldots, (X_n, Y_n), Z$ are independent. Notice that for $i = 1, 2, \ldots, n$,

$$F_i(x) = P[X_i < x] = \mu_i((-\infty, x) \times \mathbb{R})$$

$$= C_i(F(x), G(+\infty)) = C_i(F(x), 1) = F(x)$$

so the X_i's are identically distributed. Similarly, for $i = 1, 2, \ldots, n$, $G_i(y) = G(y)$ so the Y_i's are identically distributed. Also, for $i = 1, 2, \ldots, n$,

$$P[Z = i] = \mu_{n+1}(\{i\}) = p_i$$

and

$$H_i(x,y) = P[X_i < x, Y_i < y] = \mu_i((-\infty, x) \times (-\infty, y))$$

$$= C_i(F(x), G(y)) = C_i(F_i(x), G_i(y)),$$

i.e., each C_i is a copula for (X_i, Y_i). Finally, defining X^* and Y^* as in (5) we see that

$$H^*(x,y) = P[X^* < x, Y^* < y] = \sum_{j=1}^{n} P[X^* < x, Y^* < y, Z = j]$$

$$= \sum_{j=1}^{n} P[X_j < x, Y_j < y, Z = j] = \sum_{j=1}^{n} p_j\, P[X_j < x, Y_j < y]$$

$$= \sum_{j=1}^{n} p_j\, H_j(x,y) = \sum_{j=1}^{n} p_j\, C_j(F_j(x), G_j(y))$$

$$= \sum_{j=1}^{n} p_j C_j(F(x), G(y)) = H(x,y).$$

We now turn our attention to interpreting convex sums of copulas in a more general context. Specifically we assume that $(\Omega, \mathcal{A}, P)$ and $(\mathbb{R}, \mathcal{B}_1, P_1)$ are probability spaces where $\mathcal{B}_1$ is the collection of all Borel subsets of $\mathbb{R}$. We further assume that X, $Y : \Omega \to \mathbb{R}$ are random variables and that $\{C_z : z \in \mathbb{R}\}$ is a family of copulas such that the map $z \mapsto C_z(a,b)$ is Borel measurable for each (a,b) in I^2. Finally, we define a copula C via

$$C(a,b) = \int_{\mathbb{R}} C_z(a,b) dP_1(z). \tag{5.1}$$

Probabilistic Interpretation of a Convex Sum of Copulas: The copula C, defined by (5.1), is a copula for (X,Y) if and only if there is a probability space $(\Omega^*, \mathcal{A}^*, P^*)$ and random variables X^*, Y^*, Z^* defined on Ω^* such that

(a) $H = H^*$,

(b) $P^*\{\omega : Z^*(\omega) \in A\} = P_1(A)$ for all $A \in \mathcal{B}_1$, and

(c) $P^*[X^* < x, Y^* < y \mid Z^* = z] = C_z(F(x), G(y))$ for all x, y, z in $\mathbb{R}$.

<u>Proof:</u>   Suppose C is a copula for (X,Y). For each $z \in \mathbb{R}$, let

$$H_z(x,y) \equiv C_z(F(x), G(y)).$$

Since H_z is a distribution function it induces a Lebesgue-Stieltjes probability measure P_z on $(\mathbb{R}^2, \mathcal{B}_2)$ where $\mathcal{B}_2$ is the family of all Borel subsets of $\mathbb{R}^2$. The map $z \mapsto P_z([x_1, x_2) \times [y_1, y_2))$ is Borel measurable because

$$P_z([x_1, x_2) \times [y_1, y_2)) = H_z(x_2, y_2) - H_z(x_2, y_1) + H_z(x_1, y_1) - H_z(x_1, y_2).$$

Next, if A is any finite union of intervals, then the map $z \mapsto P_z(A)$ is again Borel measurable. Now let $B \in \mathcal{B}_2$. Then according to [7] there is a sequence of numbers $\alpha_1, \alpha_2, \alpha_3, \ldots \in \{-1, 1\}$ and a sequence $\{A_n\}_{n \in \mathbb{N}}$ of sets, each a finite union of intervals, such that

$$P_z(B) = \sum_{n=1}^{\infty} \alpha_n P_z(A_n).$$

For each $m \in \mathbb{N}$, the map

$$z \mapsto \sum_{n=1}^{m} \alpha_n P_z(A_n)$$

is Borel measurable. Thus by Theorem 1.5.4 in [1] we have the map $z \mapsto P_z(B)$ is Borel measurable. According to Theorem 2.6.2 in [1] there is a unique probability measure P^* on $\mathcal{A}^*$, the product σ-field $\mathcal{B}_1 \times \mathcal{B}_2$, such that

$$P^*[A \times B] = \int_A P_z(B) dP_1(z)$$

for all $A \in \mathcal{B}_1$ and all $B \in \mathcal{B}_2$. Define X^*, Y^*, Z^* on $\Omega^* = \mathbb{R} \times \mathbb{R}^2$ as follows:

$$X^*(z,(x,y)) = x,$$

$$Y^*(z,(x,y)) = y,$$

and

$$Z^*(z,(x,y)) = z.$$

Let $x, y \in \mathbb{R}$ and let $B = [-\infty,x) \times [-\infty,y)$. Then

$$H^*(x,y) = P^*[X^* < x, Y^* < y] = P^*[\mathbb{R} \times B] = \int_{\mathbb{R}} P_z(B)dP_1(z)$$

$$= \int_{\mathbb{R}} C_z(F(x), G(y))dP_1(z) = C(F(x), G(y)) = H(x,y).$$

Thus (a) is satisfied.

Next, let $A \in \mathfrak{B}_1$. Then

$$P^*(\{\omega \in \Omega^* : Z^*(\omega) \in A\}) = P^*(A \times \mathbb{R}^2)$$

$$= \int_A P_z(\mathbb{R}^2)dP_1(z) = \int_A dP_1(z) = P_1(A).$$

Therefore (b) is satisfied.

From the definition of P^* we have, for $x, y \in \mathbb{R}$ and any $A \in \mathfrak{B}_1$,

$$P^*[X^* < x, Y^* < y, Z^* \in A] = P^*(A \times B) = \int_A P_z(B)dP_1(z)$$

$$= \int_A H_z(x,y)dP_1(z)$$

$$= \int_A C_z(F(x), G(y))dP_1(z).$$

This allows us to conclude that (c) is satisfied and the proof is complete in one direction.

In the other direction, suppose there is a probability space $(\Omega^*, \mathcal{A}^*, P^*)$ and random variables X^*, Y^*, Z^* such that (a), (b) and (c) hold. Then

$$H(x,y) = H^*(x,y) = P^*[X^* < x, Y^* < y] = P^*[X^* < x, Y^* < y, Z^* \in \mathbb{R}]$$

$$= \int_{\mathbb{R}} P^*[X^* < x, Y^* < y \mid Z^* = z]dP_1(z)$$

$$= \int_{\mathbb{R}} C_z(F(x), G(y))dP_1(z) = C(F(x), G(y)).$$

Thus C is a copula for (X,Y).

In the preceding result one can, of course, replace the probability space $(\mathbb{R}, \mathfrak{B}_1, P_1)$ with a arbitrary probability space. In fact one could use a probability measure on collections of copulas.

The copula Π occupies a central position in the collection of all copulas. This statement is true from many perspectives including the following. First let C be any copula. Next, identify the left-hand and right-hand edges of I^2 and translate the mass associated with C to the right (modulo 1) by an amount t (where $0 \le t \le 1$). The resulting mass distribution will correspond to a new copula C_t given by

$$C_t(x,y) = \begin{cases} C(1+x-t,y) - C(1-t,y), & \text{if } x \leq t, \\ C(x-t,y) + y - C(1-t,y), & \text{if } x > t. \end{cases}$$ (5.2)

It is easy to prove using integration by substitution that for any copula C,

$$\int_0^1 C_t(x,y)\, dt = xy = \Pi(x,y).$$ (5.3)

Visually one can see this by imagining that the unit square endowed with the mass distribution for C is wrapped around a circular cylinder so that the left and right edges meet. If one spins the cylinder at a constant rate one will see the uniform mass distribution associated with Π.

The copula M_t, defined by (5.2) with $C=M$, is a shuffle of Min. In [6] A. W. Marshall proves the following very interesting result in which he makes use of convex sums of the copulas M_t with $0 \leq t \leq 1$. To state the result more succinctly, we define $\oplus$ to mean addition modulo 1, i.e.,

$$x \oplus y = x + y - [x+y] = x+y \pmod 1.$$

Theorem 5.1 Suppose that $P[(X,Y) \in I^2] = 1$. Then, $(X \oplus U, Y \oplus U)$ and (X,Y) have the same distribution for every random variable $U \geq 0$ independent of (X,Y) if and only if, for some distribution G with support contained in I,

$$H(x,y) = \int_0^1 M_t(x,y)\, dG(t)$$

for all (x,y) in I^2.

Using Choquet's representation theorem one can show that every copula can be represented as a convex sum (in the sense of being an integral with respect to a probability measure) of extremal copulas. Equation (5.3) shows that such a representation is not in general unique. To see this one need only replace C in (5.3) first by W and then by M and note that each W_t and each M_t is extremal.

6. Problems and Questions

In conclusion there are some interesting, worthwhile problems and questions which are related to the issues discussed in this paper and, to the best of our knowledge, remain to be answered. A more general version of the first of these appeared in [10]; it was this problem which initiated much of our work with copulas.

(1)   Give a *probabilistic interpretation* for the copula C, whenever C admits the representation:

$$C(x,y) = k^{-1}(k(x)k(y))$$

for all (x,y) in I^2, where k is a continuous and strictly increasing function from I onto I.

(2)   Whenever $C, C_1, C_2, \ldots, C_n$ are given copulas, find $p_1, p_2, \ldots, p_n$ in I such that $p_1 C_1 + p_2 C_2 + \cdots + p_n C_n$ best approximates C.

(3) Is there an <u>algorithm</u> which will always decompose a non-extremal copula into a convex sum of two different copulas?

(4) Is there an <u>algorithm</u> which permits us to start with a copula C and <u>construct</u> the representation of C as a convex sum (in the sense of being an integral with respect to a probability measure) of extremal copulas?

7. References

1. Ash, R. B. (1972) Real Analysis and Probability, Academic Press, New York and London.
2. Dall'Aglio, G. (1972) 'Fréchet classes and compatibility of distribution functions', Symposia Math. 9, 131-150.
3. Fréchet, M. (1951) 'Sur les tableau de correlation dont les marges sont donnes', Ann. Univ. Lyon *Sect. A Ser. 3* 14, 53-77.
4. Kamiński, A., Sherwood, H., and Taylor, M. D. (1988) 'Doubly stochastic measures with mass on the graph of two functions', Real Anal. Exchange 13, 253-257.
5. Kimeldorf, G. and Sampson, A. R. (1978) 'Monotone dependence', Ann. of Statist. 6, No. 4, 895-903.
6. Marshall, A. W. (1989) 'A bivariate uniform distribution', in L. J. Gleser, M. D. Perlman, S. J. Press, and A. R. Sampson (eds.) Contributions to Probability and Statistics, Springer-Verlog., New York, pp. 99-106.
7. Mikusiński, P. (1988) 'On the completion of measures', Arch. Math. (Basel) 50, 259-263.
8. Mikusiński, P., Sherwood, H., and Taylor, M. D., 'Shuffles of Min', Submitted.
9. Nelsen, R. B. (1987) 'Statistical Properties of certain doubly stochastic measures', Talk-AMS meetings, San Antonio.
10. Schweizer, B. and Sklar, A. (1983) Probabilistic Metric Spaces, North Holland, Amsterdam.
11. Seethoff, T. L. and Shifflet, R. C. (1978) 'Doubly stochastic measures with prescribed support', Z. Wahrsch. verw. Gebiete 41, 283-288.
12. Sklar, A. (1959) 'Fonctions de répartition à *n* dimensions et leurs marges', Publ. Inst. Statist. Univ. Paris 8, 229-231.
13. Vitale, R. A. (1990) 'Stochastic dependence and a class of degenerate distributions', in H. Block, A. R. Sampson, T. Savits (eds.), Topics in Statistical Dependence, IMS Lecture Notes and Monograph Series (to appear).
14. Wolff, E. F. (1977) 'Measures of dependence derived from copulas', Ph. D. Thesis, Univ. of Massachusetts, Amherst.

A NEW APPROACH TO DEPENDENCE IN MULTIVARIATE DISTRIBUTIONS

S. KOTZ
College of Business and Management
University of Maryland
College Park, MD 20742, USA

J. P. SEEGER
BBN Communications
Cambridge, MA 02140, USA

ABSTRACT. In this paper the concept of density weighting function (d.w.f.) [6] is reexamined and a new constructive approach to the generation of dependence between random variables based on this concept is proposed. A number of well known classical distributions (including the generalized Farlie-Gumbel-Morgenstern distribution) are reinterpreted and new reparametrizations are introduced. Limits of dependence explained by d.w.f.s are examined. An elementary Lemma (stated here for the case $n = 2$) which serves as a key for a number of far reaching generalizations can be formulated as follows:

> Let $h(x, y)$ be a p.d.f. on the square $[0, 1]^2$, symmetric about the line $y = x$. If the isoprobability contours of h are of the form $x - y = k (k \in [-1, 1])$, and $h(x, y)$ is a strictly monotone function of $|x - y|$, then marginal densities cannot be uniform.

Keywords and Phrases: Density Weighting Functions, Fréchet Bounds, Farlie-Gumbel-Morgenstern Distributions, Plackett Distribution, Uniform Marginals, Singular Concentration of Probability on Diagonal.

Motivation

In the last decade it has become increasingly important to consider dependence as more than an antithesis to independence, the latter being the basic concept of mathematical probability theory. As a result, several methods have been developed to impose dependence among random variables with given marginal distributions. The majority of bivariate methods are based on a well-known result due to Hoeffding [10] and Fréchet [6] which says that given any two random variables, X and Y, with respective c.d.f.s $F_1(x)$ and $F_2(y)$, the class $\Pi(F_1, F_2) = \{ H(x,y) \mid H$ is a bivariate c.d.f. with marginals $F_1(x)$ and $F_2(y) \}$ contains an upper bound, H^*, and a lower bound, H_*. These are bounds with respect to the partial ordering $\prec$, denoting stochastic dominance, i.e. if $H, H' \in \Pi(F_1, F_2)$, then $H \prec H'$ iff $H(x,y) \leq H'(x,y) \forall (x,y)$. Moreover, the so-called Fréchet bounds have general expressions in terms of F_1 and F_2, namely $H_*(x,y) = \max \{ F_1(x) + F_2(y) - 1, 0 \}$ and $H^*(x,y) = \min \{ F_1(x), F_2(y) \}$. Several classic measures of dependence, including the coefficients of correlation and association, have been shown to reach their extreme values at H^* and H_*. Hence, if $H, H' \in \Pi(F_1, F_2)$ and $H \prec H'$, then H' in some sense imposes greater positive (equivalently less negative) dependence between X and Y than does H, or according to the terminology of Tchen [27], H' is more concordant than H. In this sense H^* and H_* are extremes of positive and negative dependence respectively.

Several parametrized subsets of $\Pi(F_1, F_2)$ which are linearly ordered with respect to $\prec$ have appeared in the literature. Most classic among these is the bivariate normal c.d.f. parametrized by the correlation coefficient (see, for example, Anderson [1]). For arbitrary marginals there are the Farlie-Gumbel-Morgenstern (F.G.M.) [25], among others, and Plackett's families ([25] and [11]). For exponential marginals there is, among others, Gumbel's type I distribution [9]. Except for the bivariate normal case, all of the above were constructed according to the viewpoint that the way to impose dependence is to increase (or decrease) everywhere the independent c.d.f. $F_1(x) \times F_2(y)$ without altering the marginals, thus creating a new c.d.f. closer in value to H^* (or H_*). While this may be a valid approach, an interesting and more natural alternative exists.

1 The Density Weighting Function

It is not difficult to show that the c.d.f. H^* in $\Pi(F_1, F_2)$ corresponds to a concentration of the total probability mass on the set $\{ (x,y) \mid F_1(x) = F_2(y) \}$, and similarly H_* corresponds to concentration of mass on $\{ (x,y) \mid F_1(x) + F_2(y) = 1 \}$. If we assume, which we shall, that F_1 and F_2 are continuous and strictly increasing, then $\{ (x,y) \mid F_1(x) = F_2(y) \}$ and $\{ (x,y) \mid F_1(x) + F_2(y) = 1 \}$ become the curves $F_1(x) = F_2(y)$ and $F_1(x) + F_2(y) = 1$ respectively. It seems equally valid, now, to consider imposing dependence by altering the independent p.d.f. $f_1(x) \times f_2(y)$ in such a way to make it tend in some manner toward the singular concentrations mentioned above. (Here we are also assuming that F_1 and F_2 are absolutely continuous with respective p.d.f.s f_1 and f_2. We shall continue this assumption.) Before we can consider such a technique, however, it is necessary to define what we mean in saying the density tends toward the concentration associated with H^* or H_*. Perhaps the easiest way to motivate a definition is to examine an already well-known parametrized class of c.d.f.s whose densities behave in the desired fashion.

We are speaking here of the F.G.M. class with absolutely continuous, strictly increasing marginals F_1 and F_2 with respective densities f_1 and f_2. We shall denote by $H_\theta(x,y)$ the F.G.M. c.d.f.

$$F_1(x) \times F_2(y) [1 + \theta(1 - F_1(x))(1 - F_2(y))]$$

and by $h_\theta(x,y)$ the corresponding density which is known to be

$$f_1(x) \times f_2(y) [1 + \theta(1 - 2F_1(x))(1 - 2F_2(y))]$$

where $\theta \in [-1, 1]$. In the expression for $h_\theta(x,y)$, the independent density, $f_1(x) \times f_2(y)$, is multiplied by a function of x and y dependent on θ which we shall denote by $\phi_\theta(x, y)$. Expanding, we get

$$\phi_\theta(x, y) = 1 + \theta - 2\theta\{F_1(x) + F_2(y)\} + 4\theta F_1(x) \times F_2(y) \tag{1}$$

If we then restrict this function to a curve $F_1(x) + F_2(y) = k$, where k is some constant in $[0, 2]$, we get

$$\phi_\theta(x, y)|_{F_1(x)+F_2(y)=k} = 1 + \theta(1 - 2k) + 4\theta F_1(x) \times F_2(y)$$

The significance of this is as follows. A curve of the form $F_1(x) + F_2(y) = k$ represents y as a decreasing function of x and vice versa. Hence it intersects $F_1(x) = F_2(y)$ at one and only one point, namely $\left\{F_1^{-1}(k/2), F_2^{-1}(k/2)\right\}$. Furthermore, by simple calculus or the Lemma which follows (and which will allow us to generalize our theory to more than 2 dimensions), $F_1(x) \times F_2(y)$ restricted to $F_1(x) + F_2(y) = k$ increases towards $F_1(x) = F_2(y)$ and achieves its maximum at the point of intersection of the two curves. Therefore, when $\theta > 0 (\theta < 0)$, $\phi_\theta(x, y)|_{F_1(x)+F_2(y)=k}$ increases (decreases) towards $F_1(x) = F_2(y)$ and achieves its maximum (minimum) on $F_1(x) = F_2(y)$. Actually, for the case that θ is negative, we can look at $\phi_\theta(x, y)$ restricted to a curve of the form $1 - F_1(x) + F_2(y) = k$ on which

$$\phi_\theta(x, y)|_{1-F_1(x)+F_2(y)=k} = 1 + \theta(2k - 1) - 4\theta(1 - F_1(x))F_2(y)$$

These curves cross $F_1(x) + F_2(y) = 1$, and $(1 - F_1(x))F_2(y)|_{1-F_1(x)+F_2(y)=k}$ increases towards $F_1(x) + F_2(y) = 1$ (again by the following Lemma) and achieves its maximum on this curve. Thus, when $\theta < 0$, $\phi_\theta(x, y)|_{1-F_1(x)+F_2(y)=k}$ behaves likewise. The Lemma we have called upon here is the classic result

Lemma 1 *Let $x_1, \ldots, x_n$ be n non-negative real numbers and $a_1, \ldots, a_n$ be n positive real numbers such that $\sum_{i=1}^n a_i = 1$. Then*

$$\prod_{i=1}^n x_i^{a_i} \le \sum_{i=1}^n a_i x_i$$

with equality holding only in the case $x_1 = \cdots = x_n$.

Proof: See, for example, Rankin[26]. ∎

Perhaps better known is the special case of this Lemma in which $a_1 = a_2 = \ldots = a_n = \frac{1}{n}$. Then the statement reads that the geometric mean is dominated by the arithmetic mean, or equivalently, on the positive section of the hyperplane with equation $x_1 + x_2 + \cdots + x_n = k$, $\prod_{i=1}^n x_i$ achieves its maximum value at $x_1 = x_2 = \cdots = x_n = k/n$. For a better idea of what's happening here, consider $\phi_\theta(x, y)$ in the case $F_1(x) = x$ and $F_2(y) = y$ (i.e., they are both $U[0, 1]$ c.d.f.s) and for $\theta = 1$. $\phi_\theta(x, y)$ is a weight function with a maximal ridge along $F_1(x) = F_2(y)$. As θ increases from 0, more of the total mass is situated on and about $F_1(x) = F_2(y)$ while less is situated away from this curve. More precisely, since for $\theta > 0$

$$\phi_\theta(x, y) \begin{cases} < 1 & \text{if } F_1(x) < 1/2 \text{ and } F_2(y) > 1/2, \\ & \text{or if } F_1(x) > 1/2 \text{ and } F_2(y) < 1/2 \\ = 1 & \text{if } F_1(x) = 1/2 \text{ or } F_2(y) = 1/2 \\ > 1 & \text{if } F_1(x) < 1/2 \text{ and } F_2(y) < 1/2, \\ & \text{or if } F_1(x) > 1/2 \text{ and } F_2(y) > 1/2, \end{cases}$$

as θ increases from 0, $h_\theta(x,y)$ is increased in the quandrants through which $F_1(x) = F_2(y)$ passes and decreased in the two "outlying" quadrants. Note that at one point of $F_1(x) = F_2(y)$, namely $\left(F_1^{-1}(1/2), F_2^{-1}(1/2)\right)$, $\phi_\theta(x,y) = 1$ and does not vary with θ. This will be discussed later. The analogous situation exists in the case $\theta < 0$ with the maximal ridge of $\phi_\theta(x,y)$ occurring along $F_1(x) + F_2(y) = 1$ and the roles of the two pairs of quadrants reversed.

Although many possibilities exist for defining a tendency towards the mass concentrations of H^* and H_*, the preceding examples suggest the following approach. Alter the independent density $f_1(x) \times f_2(y)$ by multiplying it by a bivariate function $\phi(x,y)$. We'll call this function a *density weighting function* (d.w.f.). It should satisfy the following properties. First, since $f_1(x) \times f_2(y) \times \phi(x,y)$ is to be a p.d.f. with marginal p.d.f.s $f_1(x)$ and $f_2(y)$, $\phi(x,y)$ must be non-negative, and it must be true that

$$\int_{-\infty}^{\infty} \phi(x,y) \times f_1(x)dx = \int_{-\infty}^{\infty} \phi(x,y) \times f_2(y)dy = 1.$$

Next, in the case that positive dependence is to be imposed, there should be curves cutting across $F_1(x) = F_2(y)$ (graphs of y as a decreasing function of x) on which $\phi(x,y)$ increases towards $F_1(x) = F_2(y)$. (These curves should be such that every point in the support of $f_1(x) \times f_2(y)$ lies on exactly one of these curves. In other words, the cutting-across curves should partition the support $f_1(x) \times f_2(y)$.) As a result of this property, $\phi(x,y)$ should have, in some sense, a maximal ridge along $F_1(x) = F_2(y)$. In case the dependence to be imposed is negative, $F_1(x) + F_2(y) = 1$ plays the role of $F_1(x) = F_2(y)$ in the above discussion. Since, in the bivariate case, negative dependence is the opposite of positive dependence, imposing negative dependence by a d.w.f. with a trough along $F_1(x) = F_2(y)$ should also be considered. Similarly, one could conceivably create positive dependence with a d.w.f. having a trough along $F_1(x) + F_2(y) = 1$.

Having roughly defined the concept of a d.w.f. we examined four other parametrized classes of bivariate c.d.f.s each of which, like the F.G.M. class, is a subset of some $\Pi(F_1, F_2)$ and is linearly ordered with respect to $\prec$. Classes examined were Gumbel Type I Exponential, Standard Binormal, Plackett's Class (uniform representation), and Bivariate Pareto. For each class $\{H_\theta\}$ with densities $\{h_\theta\}$ we computed the d.w.f.s $\{\phi_\theta\}$ where

$$\phi_\theta(x,y) = h_\theta(x,y)/\left(f_1(x) \times f_2(y)\right)$$

Then we compared these to our proposed definition of a d.w.f. The bivariate Pareto class examined is a generalization by the authors of a specific distribution found in [11]. It is explained in Appendix I.

As an example, we consider here the Gumbel Type I Exponential class with p.d.f.

$$((1 + \theta x)(1 + \theta y) - \theta)\,e^{-x-y-\theta xy}$$

and d.w.f.

$$\phi_\theta(x,y) = \{(1 + \theta x)(1 + \theta y) - \theta\}\,e^{-\theta xy}$$

we show that $\phi_\theta(x,y)$ when restricted to $\{(1 + \theta x)(1 + \theta y) = k\}$, achieves its minimum on $\{y = x\}$. This is equivalent to showing $xy|_{\{(1+\theta x)(1+\theta y)=k\}}$ achieves its maximum on $\{y = x\}$. We shall prove this now.

Let $u = \ln(1 + \theta x)$, $v = \ln(1 + \theta y)$, and $g(t) = e^t - 1$; so that $x = \frac{g(u)}{\theta}, y = \frac{g(v)}{\theta}$, and the problem is reduced to showing that on $\{ e^{u+v} = k \}$ (this is $\{ (1 + \theta x)(1 + \theta y) = k \}$), or equivalently on $\{ u + v = \ln k \}$, $g(u)g(v) \le \left[g\left(\frac{\ln k}{2} \right) \right]^2$. This last inequality is equivalent to $\theta^2 xy \le (\sqrt{k} - 1)^2$; easily understood if one notes that $\{ y = x \} \cap \{ (1 + \theta x)(1 + \theta y) = k \} = \left\{ \left(\frac{(\sqrt{k}-1)}{\theta}, \frac{\sqrt{k}-1}{\theta} \right) \right\}$. We will be done if we show $g(u)g(v) \le (g(\frac{u+v}{2}))^2$:

$$g(u)g(v) = e^{u+v} - e^u - e^v + 1 \le e^{u+v} - 2e^{\frac{u+v}{2}} + 1$$

(because $e^u + e^v \ge 2e^{\frac{u+v}{2}}$ by Lemma 1) $= \left(e^{\frac{u+v}{2}} - 1 \right)^2 = (g(\frac{u+v}{2}))^2$.

2   Application of Lemma 1

Already having shown how the d.w.f.s of the F.G.M. and Gumbel Type I Exponential classes depend for their properties on Lemma 1, it is interesting to note that this Lemma also accounts for at least part of the behavior of the d.w.f.s of the other four classes.

In the sense of "quadrant dependence" of [21], the dependence imposed by $\theta > 0$ in the Gumbel type I class is negative. In line with the "trough" technique of imposing dependence suggested above, we've shown that $\phi_\theta(x, y)$ has in some sense a trough along $F_1(x) = F_2(y)$. Does it have a ridge along $F_1(x) + F_2(y) = 1$ which is $e^{-x} + e^{-y} = 1$? One approach to this question is to consider, for example, the equation

$$\frac{d}{dy} \phi_\theta(x, y) = 0$$

which results in

$$y = -\frac{1}{\theta} + \frac{1 + 2\theta x}{(1 + \theta x)\theta x},$$

a decreasing convex function of x whose graph has as asymptotes $x = 0$ and $y = -\frac{1}{\theta}$. It is not coincident with $e^{-x} + e^{-y} = 1$. One could also take the directional derivative of $\phi_\theta(x, y)$ along a line normal to $e^{-x} + e^{-y} = 1$ and find its zeros.

The d.w.f. of the bivariate Pareto distribution, when $\theta = 1$, may be written,

$$\frac{(\alpha_2 x \alpha_1 y)^{a+1}(a + 1)}{a(\alpha_1 \alpha_2)^a(\alpha_2 x + \alpha_1 y - \alpha_1 \alpha_2)^{a+2}}.$$

Also, when $\theta = 1$, the partitioning curves take the form $\alpha_2 x + \alpha_1 y = k, k \in (2\alpha_1\alpha_2, \infty)$. On these lines $\phi_1(x, y)$ varies as $(\alpha_2 x \alpha_1 y)^{a+1}$, and hence, by Lemma 1, attains its maximum at $\alpha_2 x = \alpha_1 y$ which is the equation of $F_1(x) = F_2(y)$. For $\theta \in (0, 1)$, the connection with Lemma 1 is more subtle. It suffices to notice that along $(1-\theta)xy + \theta(\alpha_2 x + \alpha_1 y) = k, \phi_\theta(x, y)$ varies only as $(xy)^{a+1}$ which, on this curve, attains its maximum at the intersection with $\alpha_2 x = \alpha_1 y$.

Along $x^2 + y^2 = k$ the d.w.f. of the standard binormal class varies only as $\exp(\frac{\theta}{1-\theta^2} xy)$. By Lemma 1, on $x^2 + y^2 = k$, $x^2 y^2$ attains its maximum at $x^2 = y^2$. Therefore, on $x^2 + y^2 = k$, xy attains its maximum at $x = y$ (two points) and its minimum (the negative of its maximum) at $x = -y$ (two points). That is, for $\theta > 0, \phi_\theta(x, y)$ has a ridge along

$y = x$ and a trough along $y = -x$ (resulting in a saddle point at $(0,0)$). We get the opposite situation when $\theta < 0$. Another example of a binormal distribution with parametrized dependence is constructed in Appendix 2.

Finally, along $x + y = k(k \in [0,2])$, letting $xy = z$, the d.w.f. of the uniform version of Plackett's distribution takes the form

$$\frac{\theta(\theta - 1)(k - 2z) + \theta}{\left[((\theta - 1)k + 1)^2 - 4\theta(\theta - 1)z\right]^{\frac{3}{2}}},$$

an increasing function of z when $\theta > 1$. Since z attains its maximum at $y = x$, $\phi_\theta(x,y)$ has a ridge along $y = x$ when $\theta > 1$. It can be shown in a similar way that $\phi_\theta(x,y)$ has a ridge along $x + y = 1$ when $\theta < 1$. (In this case the role of the partitioning curves is played by lines of the form $y = x + k, k \in [-1,1]$.)

3   Limits of Dependence

By a result due to Hoeffding (see [10], [21], or [28]), and known by some as Hoeffding's Lemma, one obtains the following

Lemma 2 *If $H, H' \in \Pi(F_1, F_2)$, and $H' \succ H$, then (provided the correlations exist)* $\text{corr}_{H'}(X,Y) \geq \text{corr}_H(X,Y)$. *Consequently, for any $H \in \Pi(F_1, F_2)$, $\text{corr}_{H_*}(X,Y) \leq \text{corr}_H(X,Y) \leq \text{corr}_{H^*}(X,Y)$.*

Therefore, since each of our five classes is linearly ordered by $\prec$, it is also linearly ordered by correlation coefficients. Hence, $\text{corr}_{H^*}(X,Y) - \text{corr}_{H_\theta}(X,Y)$ or $\text{corr}_{H_*}(X,Y) - \text{corr}_{H_\theta}(X,Y)$ are valid measures of the effectiveness of $\phi_\theta(x,y)$ when the dependence being imposed is positive or negative respectively. Whenever F_1 and F_2 differ at most by scale and/or location parameters, then $F_1(x) = F_2(y)$ is a straight line causing corr $_{H^*}(X,Y) = 1$. Hence, when the given marginals of an F.G.M. class are so related, then $1 - \text{corr}_{H_\theta}(X,Y)$(for $\theta > 0$) measures the effectiveness of $\phi_\theta(x,y)$. If, in addition, $F_1(x) + F_2(y) = 1$ is a straight line, then corr $_{H_*}(X,Y) = -1$, so that, for $\theta < 0, -1 - \text{corr}_{H_\theta}(X,Y)$ is a measure of the strength of $\phi_\theta(x,y)$. Analogous statements may be made for the standard binormal and uniform Plackett's families both of which satisfy corr $_{H^*}(X,Y) = 1$ and corr $_{H_*}(X,Y) = -1$. The bivariate Pareto d.w.f. imposes only positive dependence, and for this class F_1 and F_2 differ only by a scale parameter. Therefore, $1 - \text{corr}_{H_\theta}(X,Y)$ measures the strength of the Pareto d.w.f. The Gumbel type I d.w.f. imposes only negative dependence, but here $F_1(x) + F_2(y) = 1$ is $e^{-x} + e^{-y} = 1$, not a straight line. However, by numerical methods, corr $_{H_*}(X,Y)$ is found to be approximately $-.645$, so comparison of $\text{corr}_{H_\theta}(X,Y)$ with this value will indicate the strength of $\phi_\theta(x,y)$. Table 1 gives the values of $\text{corr}_{H_\theta}(X,Y)$ for extreme values of θ for each of the five classes considered previously. The F.G.M. class is listed with three different sets of marginals.

Table 1 shows that the d.w.f.s of the F.G.M. classes, the Gumbel type I class, and the bivariate Pareto class are relatively inefficient even at their strongest. This can be explained by the fact that the action of these d.w.f.s is primarily dependent on Lemma 1. In each of the three families, $\phi_\theta(x,y)$ is everywhere finite for every value of θ. In order to produce a concentration of mass on $F_1(x) = F_2(y)$, for example, we must have a d.w.f., $\phi_\theta(x,y)$, such

Distribution	θ	$\mathrm{corr}_{H_\theta}(X,Y)$	Reference
F.G.M. with uniform	1	1/3	[13]
[0, 1] marginals	−1	−1/3	[13]
F.G.M. with standard	1	$1/\pi$	[13]
normal marginals	−1	$-1/\pi$	[13]
F.G.M. with standard	1	.25	[13]
exponential marginals	−1	−.25	[13]
Gumbel type I	1	−.40365	[11]
exponential			
Bivariate Pareto	1	a^{-1} $(a > 2)$	[11]
Standard binormal	1	1	[1]
	−1	−1	
Uniform representation	∞	1	[17],
of Plackett's family	0	−1	[18]

Table 1

that as θ approaches its limiting value (upper or lower depending on the family),

$$\phi_\theta(x,y) \rightarrow \begin{cases} 0 & \text{when } F_1(x) \neq F_2(y) \\ \infty & \text{when } F_1(x) = F_2(y) \end{cases}$$

It is also noteworthy that in the F.G.M. case, the contour $\phi_\theta(x,y) = 1$ is fixed, regardless of θ, i.e., the density will never be changed along

$$\{\, (x,y) | \, (1 - 2F_1(x))\,(1 - 2F_2(y)) = 0 \,\},$$

and this set even includes one point of $F_1(x) = F_2(y)$. This was hinted at in Section 1. This partially explains why the F.G.M. families cannot demonstrate as strong a dependence as can the Gumbel type I and bivariate Pareto classes. Incidentally, the above reasoning concerning the F.G.M. d.w.f. applies even when $F_1(x) = F_2(y)$ is not a straight line.

In contrast to the other classes, the binormal and Plackett's families each encompass the entire range of $\mathrm{corr}(X,Y)$. These two families, in addition to containing the independent distributions of their respective classes, also contain both upper and lower Fréchet bounds. In the binormal case we have

$$\lim_{\theta \to 1} \phi_\theta(x,y) = \begin{cases} 0 & \text{if } y \neq x \\ \infty & \text{if } y = x \end{cases} \quad \text{and} \quad \lim_{\theta \to -1} \phi_\theta(x,y) = \begin{cases} 0 & \text{if } y \neq -x \\ \infty & \text{if } y = -x. \end{cases}$$

For Plackett's class we have

$$\lim_{\theta \to \infty} \phi_\theta(x,y) = \begin{cases} 0 & \text{if } y \neq x \\ \infty & \text{if } y = x \end{cases} \quad \text{and} \quad \lim_{\theta \to 0} \phi_\theta(x,y) = \begin{cases} 0 & \text{if } y \neq -x \\ \infty & \text{if } y = -x. \end{cases}$$

These same conditions are satisfied by Frank's family of bivariate distributions investigated by Genest [7].

4   The Shapes of D.W.F.s

It can be shown that for positive θ the uniform F.G.M. d.w.f. is saddle-shaped with its saddle point at the intersection of $F_1(x) = F_2(y)$ with $F_1(x) + F_2(y) = 1$. It's not difficult to show that the binormal d.w.f. fits the same description while Plackett's uniform d.w.f. behaves similarly in a neighborhood of $F_1(x) = F_2(y)$. In other words, for these classes, although $\phi_\theta(x,y)$ for $\theta > 0$ ($\theta > 1$ in Plackett's case) increases toward $F_1(x) = F_2(y)$ on every cross contour, it gives more weight to regions where x and y are both large or both small than it does to the region surrounding $(F_1^{-1}(\frac{1}{2}), F_2^{-1}(\frac{1}{2}))$. In fact, the univariate function $\phi_\theta\left(x, F_2^{-1}(F_1(x))\right)$ has a minimum at this point. This might lead one to believe that these d.w.f.s are not causing the p.d.f. to approach a concentration on $F_1(x) = F_2(y)$ as efficiently as possible. For example, at points of $y = x$ away from the origin, the binormal d.w.f. with positive θ is

$$(1 - \theta^2)^{-\frac{1}{2}} \times \exp\left(\theta x^2/(1 + \theta)\right),$$

a product of two increasing functions of θ, while at $(0,0)$, it is $(1 - \theta^2)^{-\frac{1}{2}}$, the factor which approaches ∞ as $\theta \to 1$. Why can't there be a d.w.f. which weights the p.d.f. with equal intensity along the entire length of $F_1(x) = F_2(y)$? Some insight into the question is gained from the following proposition and the example presented in Appendix 2.

Proposition 1 *Let $h(x,y)$ be a p.d.f. on the square, $[0,1]^2$, symmetric about the line $y = x$. If the isoprobability contours of h are of the form $x - y = k$ ($k \in [-1,1]$) and $h(x,y)$ is a strictly monotone function of $|x - y|$, then the marginal densities cannot be uniform.*

Proof: We prove the proposition for the case that $h(x,y)$ is a strictly decreasing function of $|x - y|$. The other case is proved similarly.

Suppose the marginal densities are uniform. Then, in particular,

$$\int_0^1 h(x,y)dy = 1 \qquad \forall x \in [0,1]$$

. Consider $\int_0^1 h(\frac{1}{2},y)dy$ and $\int_0^1 h(a,y)dy$ for some $a < \frac{1}{2}$. By hypothesis, $\int_0^{a+\frac{1}{2}} h(a,y)dy = \int_{\frac{1}{2}-a}^1 h(\frac{1}{2},y)dy$, and $\int_{a+\frac{1}{2}}^1 h(a,y)dy < \int_0^{\frac{1}{2}-a} h(\frac{1}{2},y)dy$. (Each point of $\{a\} \times \left(a + \frac{1}{2}, 1\right]$, $(a, a+\frac{1}{2}+d)$, is further from $y = x$ than the corresponding point of $\{\frac{1}{2}\} \times \left(0, \frac{1}{2} - a\right]$, $\left(\frac{1}{2}, \frac{1}{2} - a - d\right)$ for any $d > 0$.) Hence, $\int_0^1 h(a,y)dy < \int_0^1 h(\frac{1}{2},y)dy$, a contradiction. ∎

Note that the distance from a point (x_0, y_0) to the line $y = x$ is $2^{-\frac{1}{2}}|x_0 - y_0|$. Therefore, Proposition 1 actually says that a p.d.f. on $[0,1]^2$ with uniform marginals and symmetric in x and y cannot be a monotone function of the distance from $y = x$. The importance of this proposition is magnified by the following result.

Proposition 2 *It is well known (see, for example, [17], [18], [22] or [24]) that if $H \in \Pi(F_1, F_2)$, and if we define $K(u,v) = H\left(F_1^{-1}(G_1(u)), F_2^{-1}(G_2(v))\right)$ then $K \in \Pi(G_1, G_2)$.*

Distribution	$\phi_\theta\left(F_1^{-1}(u), F_2^{-1}(v)\right)$
F.G.M.	$1 + \theta(1 - 2u)(1 - 2v)$
Gumbel Type 1	$((1 - \theta\ln(1 - u))(1 - \theta\ln(1 - v)) - \theta)\ \times$ $\exp\left(-\theta\ln(1 - u)\ln(1 - v)\right)$
Bivariate Pareto	$\dfrac{((1-u)(1-v))^{-\frac{a+1}{a}}\left[a\left((1-\theta)(1-u)^{-\frac{1}{a}}+\theta\right)\left((1-\theta)(1-v)^{-\frac{1}{a}}+\theta\right)+\theta\right]}{a\left[((1-u)(1-v))^{-\frac{1}{a}}-\theta\left((1-u)^{-\frac{1}{a}}-1\right)\left((1-v)^{-\frac{1}{a}}-1\right)\right]^{a+2}}$
Standard binormal	$(1 - \theta^2)^{-\frac{1}{2}}\exp\left[-\left(2(1 - \theta^2)\right)^{-1}\ \times\right.$ $\left. \left(\theta^2\left(\left(\Phi^{-1}(u)\right)^2 + \left(\Phi^{-1}(v)\right)^2\right) - 2\theta\Phi^{-1}(u)\Phi^{-1}(v)\right)\right]$ where $\Phi()$ is the univariate standard normal C.D.F.
Plackett's	$\dfrac{\theta(\theta-1)(x+y-2xy)+\theta}{\left[((\theta-1)(x+y)+1)^2 - 4\theta(\theta-1)xy\right]^{\frac{3}{2}}}$

Table 2

Let F_1, F_2, G_1 and G_2 be absolutely continuous with respective p.d.f.s f_1, f_2, g_1 and g_2, and let H have p.d.f. $f_1(x)f_2(y) \times \phi(x, y)$. Then the p.d.f. of K is

$$g_1(u)g_2(v) \times \phi\left(F_1^{-1}(G_1(u)), F_2^{-1}(G_2(v))\right),$$

i.e. the structure of the d.w.f. is preserved under translation.

Proof: Simply compute $\frac{d^2}{dudv}K(u, v)$. ∎

Corollary 1 For any $H \in \Pi(F_1, f_2)$ with d.w.f. $\phi(x, y), \phi\left(F_1^{-1}(u), F_2^{-1}(v)\right)$ is a p.d.f. on $[0, 1]^2$ with uniform marginals.

Combining this corollary with Proposition 1 yields

Theorem 1 If $H \in \Pi(F_1, F_2)$ with d.w.f. $\phi(x, y)$, and if $\phi\left(F_1^{-1}(u), F_2^{-1}(v)\right)$ is symmetric in u and v, then it cannot be a strictly monotone function of $|u - v|$ alone.

Although more work is being done along these lines, it seems likely that the saddle shapes of the three different d.w.f.s are necessitated by similar (if not the same) basic principles. The asymmetry (about the origin) of the bivariate Pareto and Gumbel type I distributions make them difficult to analyze with respect to this theory. However, they're included in Table 2 showing $\phi_\theta\left(F_1^{-1}(u), F_2^{-1}(v)\right)$ for each of our five classes. See also the family of densities presented in Appendix 2 and Frank's family [7].

5   A Multivariate Extension

Much of the theory of bivariate dependence presents considerable difficulty when one attempts to generalize it to more than two dimensions. We have given in this paper one approach to the problem of imposing dependence between two random variables, X and Y, which is easily extendable to more than two random variables, $X_1, \ldots, X_n$. Let $\Pi(F_1, \ldots, F_n)$ denote the class of n-variate c.d.f.s with univariate marginals $F_1, \ldots, F_n$. Then $H^*(x_1, \ldots, x_n) = \min\{F_1(x_1), \ldots, F_n(x_n)\}$ is a generalization of the Fréchet upper bound. This result is due to [14] and [15]. As with $H^*(x, y)$, $H^*(x_1, \ldots, x_n)$ corresponds to a concentration of mass on the curve $F_1(x_1) = \ldots = F_n(x_n)$ and hence, to extreme positive dependence. (See Dykstra et. al. [4] for multivariate generalizations of the dependence concepts of Lehmann [21]). Kemp [16] and one of the authors of this paper (J.P.S.) have proved that $H_*(x, y)$ has no multivariate generalization as a lower bound of $\Pi(F_1, \ldots, F_n)$, but it has generalizations in the sense that mass can be concentrated on other "diagonals." For example, for $n = 3$, mass can be concentrated on $1 - F_1(x_1) = F_2(x_2) = F_3(x_3)$, $F_1(x_1) = 1 - F_2(x_2) = F_3(x_3)$, or $F_1(x_1) = F_2(x_2) = 1 - F_3(x_3)$. However, let us use the concentration associated with $H^*(x_1, \ldots, x_n)$ as an example for multivariate generalization of our dependence theory.

Recall that Lemma 1 in Section 1 was presented in general form for n real variables $x_1, \ldots, x_n$ and was followed by the alternative interpretation that on the positive orthant section of the hyperplane $x_1 + \ldots + x_n = k$, where k is some positive constant, $\prod_{i=1}^{n} x_i$ achieves its maximum at the point $(\frac{k}{n}, \ldots, \frac{k}{n})$. Expression (1) in Section 1 shows how this Lemma comes into play in the bivariate dependence imposed by an F.G.M. distribution. The connection is even clearer if we let $F_1(x) = x$ and $F_2(y) = y$ in Expression (1) obtaining

$$\phi_\theta(x, y) = 1 + \theta\left(1 - 2(x + y) + 4xy\right). \tag{2}$$

With the general form of Lemma 1 in mind, we present the following generalization of (2), a d.w.f. and hence a p.d.f. for $\Pi(U, U, U)$ where U is the uniform [0,1] c.d.f.

$$\phi_\theta(x, y, z) = 1 + \theta\left(1 - (x + y + z) + 4xyz\right) \qquad (|\theta| \leq 1) \tag{3}$$

For $\Pi(U, U, U)$, $F_1(x) = F_2(y) = F_3(z)$ is the straight line $x = y = z$. The points of $[0,1]^3$ may be partitioned into the plane sections $x + y + z = k$, where $k \in [0,3]$, each of which is pierced by $x = y = z$ at one point, $(\frac{k}{3}, \frac{k}{3}, \frac{k}{3})$. On each of these sections xyz achieves its maximum at this point of intersection. Hence, $\theta > 0$ weights the density toward $x = y = z$. For interest's sake, the c.d.f. associated with Equation 3 is

$$H_\theta(x, y, z) = xyz\left\{1 + \theta(1 - \frac{1}{2}(x + y + z) + \frac{1}{2}(xyz)\right\}. \tag{4}$$

The quantity in braces is equal to

$$1 + \frac{\theta}{2}\left((1 - x)(1 - y) + (1 - x)(1 - z) + (1 - y)(1 - z) - (1 - x)(1 - y)(1 - z)\right) \tag{5}$$

and thus $H_\theta(x, y, z)$ is recognizable as a 3-dimensional F.G.M. c.d.f. as defined in [12]. The bivariate marginal c.d.f. of Equation 4 is $xy\left\{1 + \frac{\theta}{2}(1 - x)(1 - y)\right\}$ of the same form as the c.d.f. associated with Equation 2 but with a weaker limit of dependence. Finally, Equation 2 may be generalized for any dimension n as

$$\phi_\theta(x_1, \ldots, x_n) = 1 + \frac{\theta}{n-1}\left(n - 1 - 2\sum_{i=1}^{n} x_i + 2^n \prod_{i=1}^{n} x_i\right) \quad (|\theta| \leq 1) \tag{6}$$

with c.d.f.

$$H_\theta(x_1,\ldots,x_n) = \prod_{i=1}^n x_i \left\{ 1 + \frac{\theta}{n-1} \left(n - 1 - \sum_{i=1}^n x_i + \prod_{i=1}^n x_i \right) \right\} \tag{7}$$

which can be written

$$H_\theta(x_1,\ldots,x_n) = \prod_{i=1}^n x_i \left\{ 1 + \frac{\theta}{n-1} \left(\sum_{k=2}^n (-1)^k \sum_{1 \le i_1 < \ldots < i_k \le n} \prod_{j=1}^k (1 - x_{i_j}) \right) \right\} \tag{8}$$

and has $(n-1)$-dimensional marginal c.d.f.s of the same form. Of course, Equation 7 and Equation 8 may be generalized via translation to distributions with any given univariate marginals $F_1,\ldots,F_n$. Simply substitute $F_i(x_i)$ for $x_i (i = 1, 2, \ldots, n)$ in Equation 7 and Equation 8. If $F_1,\ldots,F_n$ are absolutely continuous with respective p.d.f.s $f_1,\ldots,f_n$, the generalization of Equation 6 is formed by substituting $F_i(x_i)$ for x_i and multiplying the entire expression by $\prod_{i=1}^n f_i(x_i)$.

6 Another Application

Knowing the process by which the bivariate F.G.M. d.w.f. weights the density towards $F_1(x) = F_2(y)$ allows us to improve its efficiency in this respect. Again let us consider F.G.M. with uniform $[0,1]$ marginals. Recall that $\phi_\theta(x, y) = 1 + \theta(1 - 2x)(1 - 2y)$. We could increase the weighting effect of $W = (1 - 2x)(1 - 2y)$ if we could increase the absolute value of W when $W \ne 0$ without changing its sign or its symmetries (to preserve the marginals, we would still need $\int_0^1 W \, dx = \int_0^1 W \, dy = 0$). This can be accomplished by taking odd roots of W. Hence, we define

$$\phi_{\theta,n}(x, y) = 1 + \theta(1 - 2x)^{1/(2n-1)}(1 - 2y)^{1/(2n-1)} \qquad (|\theta| \le 1) \quad n = 1, 2, \ldots \tag{9}$$

The associated c.d.f. is

$$H_{\theta,n}(x, y) = xy + \theta \left(\frac{2n-1}{4n} \right)^2 \left(1 - (1 - 2x)^{\frac{2n}{2n-1}} \right) \left(1 - (1 - 2y)^{\frac{2n}{2n-1}} \right), \tag{10}$$

and this is an F.G.M. c.d.f. in the extended sense of [5]. Under this distribution

$$\text{corr}_{H_{\theta,n}}(X, Y) = 3\theta \frac{(2n-1)^2}{(4n-1)^2}, \text{ and } \lim_{n \to \infty} \text{corr}_{H_{\theta,n}}(X, Y) = \frac{3\theta}{4}.$$

The form of the limiting density is as follows:

$$h_\theta(x, y) = \begin{cases} 1 + \theta & \text{for } 0 < x < \frac{1}{2}, 0 < y < \frac{1}{2} \text{ and } \frac{1}{2} < x < 1, \frac{1}{2} < y < 1 \\ 1 - \theta & \text{for } 0 < x < \frac{1}{2}, \frac{1}{2} < y < 1, \text{ and } \frac{1}{2} < x < 1, 0 < y < \frac{1}{2} \end{cases}$$

Thus, we have enabled the limits of $\text{corr}(X, Y)$ to approach $\frac{3}{4}$ and $-\frac{3}{4}$, a considerable improvement. As we did in Section 5 with Equation 7, Equation 10 may be generalized via translation to a distribution with any given marginals.

7 Conclusion

We believe that in the material above, we have put forward a new approach to multivariate dependence which merits further research. Two interrelated problems present themselves. Can the d.w.f. be characterized beyond negative results like Theorem 1? What, if any, methods of constructing new d.w.f.s exist? We might also ask whether a d.w.f. can be derived from a given physical model of dependence. These questions arise from viewing the d.w.f. as a technique for constructing dependent distributions. One might also utilize the concept as a measure of dependence in already existing distributions. For example, the derivative of the d.w.f. with respect to the distance from $F_1(x_1) = \ldots = F_n(x_n)$ might be useful in comparing the dependence properties of two elements of $\Pi(F_1, \ldots, F_n)$ and might even be estimable. Finally, the corollary to Proposition 2 lends more justification to the convenient practice of studying different types of dependence as they appear in the theater of $\Pi(U, \ldots, U)$, distributions on $[0, 1]^n$ with uniform marginals.

Appendix I

Introducing a dependence-controlling parameter into the bivariate Pareto distribution given in [11] is easier from inspection of the survival function, $Pr[X > x, Y > y]$, which we shall denote by $\overline{H}(x, y)$. From [11] we find that

$$\overline{H}(x, y) = \left(\frac{\alpha_1 \alpha_2}{\alpha_2 x + \alpha_1 y - \alpha_1 \alpha_2} \right)^a \qquad (x > \alpha_1, y > \alpha_2) \tag{11}$$

while the marginal survival functions are

$$\overline{F}_1(x) = \left(\frac{\alpha_1}{x} \right)^a \text{ and } \overline{F}_2(y) = \left(\frac{\alpha_2}{y} \right)^a$$

giving an independent bivariate survival function

$$\overline{F}_1(x)\overline{F}_2(y) = \left(\frac{\alpha_1 \alpha_2}{xy} \right)^a . \tag{12}$$

Now, notice that both Equation 11 and Equation 12 can be written in the form

$$\overline{H}_\theta(x, y) = \left(\frac{\alpha_1 \alpha_2}{xy - \theta(x - \alpha_1)(y - \alpha_2)} \right)^a \qquad (\theta \in [0, 1]) \tag{13}$$

where the case of $\theta = 1$ corresponds to Equation 11 and that of $\theta = 0$ corresponds to Equation 12. The associated p.d.f. then, is

$$h_\theta(x, y) = \frac{a(\alpha_1 \alpha_2)^a \left[a\left((1 - \theta)x + \theta\alpha_1 \right)\left((1 - \theta)y + \theta\alpha_2 \right) + \theta\alpha_1\alpha_2 \right]}{(xy - \theta(x - \alpha_1)(y - \alpha_2))^{a+2}} \tag{14}$$

From Equation 13 it is easy to see that the resulting class, $\{ H_\theta \}$, is linearly ordered with respect to $\prec$.

Appendix II: An Example of a Bivariate Normal Density with Almost Ideal Positive Dependence

We define a family:

$$f_\theta(x,y) = 1 + h_\theta(x - y)$$

of smooth densities on $[0,1]^2$ with uniform marginals as follows.

h_θ is a smooth function on $[-1,1]$, periodic with period 1, with $h_\theta(0) = \theta$ and $\int_0^1 h_\theta(t)dt = 0$. h_θ is chosen to smooth the functions (on $[0,1]$):

$$\psi_\theta(x) = \begin{cases} \theta - (\theta + 1)^2 x & \text{if } 0 \le x \le \frac{1}{\theta+1} \\ -1 & \text{if } \frac{1}{\theta+1} \le x \le \frac{\theta}{\theta+1} \\ \theta - (\theta + 1)^2(1 - x) & \text{if } \frac{\theta}{\theta+1} \le x \le 1 \end{cases}$$

Note that $f_\theta(x,y)$ is not a monotone function of $(x - y)$ (cf Proposition 1). It can be shown that

$$\int_0^1 f_\theta(x,y)\,dy = \int_0^1 (1 + h_\theta(x - y))\,dy = 1 + \int_0^1 h_\theta(x - y)\,dy = 1$$

Furthermore, one can prove that as $\theta \to +\infty$, $f_\theta(x,y) \to 0$ uniformly for $\varepsilon < |x - y| < 1 - \varepsilon$, and

$$f_\theta(x,y) \to \infty \text{ if } x = y \text{ or } (x,y) = (0,1) \text{ or } (x,y) = (1,0).$$

Then one can find that

$$\lim_{\theta \to \infty} P_\theta(|x - y| \le \varepsilon) = 2 \lim_{\theta \to \infty} \int_0^\varepsilon (1 + h_\theta(v))(1 - v)dv = 1$$

Now, consider the transformation of f_θ to the class:

$$g_\theta(x,y) = \frac{1}{2\pi} e^{-\frac{1}{2}(x^2 + y^2)} \times (1 + h_\theta(\Phi(x) - \Phi(y)))$$

where, here, Φ is the normal distribution function. It can be shown that g_θ is a density function on R^2 with standard normal marginal densities. It can further be shown that

$$\lim_{\theta \to \infty} g_\theta(x,y) = \begin{cases} \infty & \text{if } x = y \\ 0 & \text{if } x \ne y \end{cases}$$

Note, however, that $g_\theta(x,y) \to \infty$ for $x \to -\infty$ and $y \to +\infty$ or $x \to +\infty$ and $y \to -\infty$. The nonuniformity of convergence off the line $x = y$ has not been completely eliminated. This is due to the requirement that $f_\theta(x,y)$ has uniform marginals.

References

[1] Anderson, T.W., *An Introduction to Multivariate Statistical Analysis*, John Wiley and Sons, Inc., New York, 1958.

[2] Dall'Aglio, G., "Les Fonctions Extrêmes de la Classes de Fréchet à Trois Dimensions," *Publ. Inst. Stat. Univ. Paris*, 9, pp. 175-188, 1960.

[3] Dall'Aglio, G., "Fréchet Classes and Compatibility of Distribution Functions" *Symp. Math*, **9**, pp. 131-150, 1972.

[4] Dykstra, R.L., J.E. Hewett, and W.A. Thompson, "Events Which Are Almost Independent," *Ann. Stats.*, **1**, pp. 674-681, 1973.

[5] Farlie, D.J.G., "The Performance of Some Correlation Coefficients for a General Bivariate Distribution," *Biometrika*, **47**, pp. 307-323, 1960.

[6] Fréchet, M., "Sur les Tableaux de Corrélation dont les Marges sont Données," *Ann. Univ. Lyon*, Sect. A, **14**, pp. 53-77, 1951.

[7] Genest, C., "Frank's Family of Bivariate Distributions," *Biometrika*, **74**, pp. 549-555, 1987.

[8] Gumbel, E.J., "Distributions à Plusieurs Variables dont les Marges sont Données," (with remarks by M. Fréchet), *Comptes Rendus de l'Académie des Sciences*, **246**, pp. 2717-2720, Paris, 1960.

[9] Gumbel, E.J., "Bivariate Exponential Distributions," *JASA*, **5**, pp. 698-707, 1960.

[10] Hoeffding, W., "Masstabinvariante Korrelationstheorie," *Schriften des mathematischen Instituts und des Instituts für Angewandte Mathematik der Universitat Berlin*, **5**, pp. 179-233, 1940.

[11] Johnson, N.L. and S. Kotz, *Distributions in Statistics: Continuous Multivariate Distributions*, John Wiley and Sons, Inc., New York, 1972.

[12] Johnson, N.L. and S. Kotz, "On Some Generalized Farlie-Gumbel-Morgenstern Distributions," *Comm. Stat.*, **4**, pp. 415-427, 1975.

[13] Johnson, N.L. and S. Kotz, "On Some Generalized Farlie-Gumbel-Morgenstern Distributions II: Regression, Correlation, and Further Generaliztions," *Comm. Stat. Theor. Meth.*, **A6(6)**, pp. 485-496, 1977.

[14] Kemp, J.F., Jr., "Advanced Problem #5894," *Am. Math Monthly*, **80**, p. 83, 1973.

[15] Kemp, J.F., Jr., "Solution to Advanced Problem #5894 by C.R. Blyth, B.C. Arnold, et. al." *Am. Math. Monthly*, **81**, p. 413, 1974.

[16] Kemp, J.F., Jr., "Advanced Problem #6115," *Am. Math. Monthly*, **83**, p. 748, 1976.

[17] Kimeldorf, G. and A. Sampson, "One-parameter Families of Bivariate Distributions with Fixed Marginals," *Comm. Stat.* , **4**, pp. 293-301, 1975.

[18] Kimeldorf, G. and A. Sampson, "Uniform Representations of Bivariate Distributions," *Comm. Stat.* , **4**, pp. 617-627, 1975.

[19] Kotz, S. and N.L. Johnson, "Dependence Properties of Iterated Generalized FGM Distributions," *Comptes Rendus de l'Académie des Sciences*, Sect. A, **285**, pp. 277-280, Paris, 1977.

[20] Kotz, S., "Dependence Concepts and their Application in Probabilistic Modelling," *Proc. 34th Annual Conference of American Society for Quality Control*, pp. 245-251, Atlanta, Georgia, 1980.

[21] Lehmann, E.L., "Some Concepts of Dependence," *Ann. Math. Stat.* , **37**, pp. 1137-1153, 1966.

[22] Mardia, K.V., *Families of Bivariate Distributions*, Hafner Publishing Co., Darien, Connecticut, 1970.

[23] Morgenstern, D., "Einfache Beispiele Zweidimensionaler Verteilungen," *Mitteilungsblatt für mathematische Statistik*, pp. 234-235, 1956.

[24] Nataf, A., "Détermination des Distributions de Probabilitiés dont les Marges sont Données," *Comptes Rendus de l'Académie des Sciences*, **255**, pp. 42-43, Paris, 1962.

[25] Plackett, F.L., "A Class of Bivariate Distributions," *JASA*, **60**, pp. 516-522, 1965.

[26] Rankin, R.A., *An Introduction to Mathematical Analysis,*, McMillan, New York, p. 224, 1963.

[27] Tchen, A., "Inequalities for Distributions with Given Marginals," *Ann. Prob*, **8**, pp. 814-827, 1980.

[28] Whitt, W., "Bivariate Distributions with Given Marginals," *Ann. Stats.*, **4**, pp. 1280-1289, 1976.

A FAMILY OF PARTIAL ORDERINGS FOR POSITIVE DEPENDENCE AMONG FIXED MARGINAL BIVARIATE DISTRIBUTIONS.

M. H. METRY AND A. R. SAMPSON [1]
University of Pittsburgh
Department of Mathematics and Statistics
Pittsburgh, Pennsylvania 15260, USA

ABSTRACT. This paper introduces a systematic and intuitive approach to generating a family of partial orderings, for positive dependence, on bivariate distributions with fixed-marginals. This family includes the more concordant and the more TP_2 orderings, as well as some existing orderings. Positive dependence ordering (PDO) properties are discussed for these orderings. Also some implications and equivalences among these orderings are established.

1. Introduction and Summary

Orderings among bivariate random variables for positive dependence is a relatively recent area of research. The basic goal of these orderings is to be able to meaningfully compare two bivariate distributions with the same x-marginals and the same y-marginals in order to see if one distribution has more positive dependence in a certain sense than the other distribution. Orderings of this nature have provided approaches: (i) to stochastically compare certain statistics' distributions (e.g., Schriever (1985), (1987a), (1987b), Tchen (1980)); (ii) to evaluate the meaningfulness of parameters in one-parameter families of fixed marginal distributions (e.g., Yanagimoto (1990)), Kimeldorf and Sampson (1987), Ahmed, Langberg, Léon and Proschan (1979)); (iii) to induce orderings on permutations via orderings for dependence on bivariate empirical c.d.f.'s (Block, Chhetry, Fang and Sampson (1991), Metry and Sampson (1988)); (iv) to provide nonparametric statistical results (Yanagimoto and Okamoto (1969)); and (v) to study the effects of random censoring on information (Hollander, Proschan and Sconing (1990)).

One widely studied ordering on bivariate distributions is the more concordant ordering (Tchen (1980)), also called the more positively quadrant dependent ordering (Ahmed, Langberg, Léon and, Proschan (1979)) and the larger quadrant dependence ordering (Yanagimoto and Okamoto (1969)). Other orderings of interest include more row regression dependent and more column regression dependent (Schriever (1985) and Yanagimoto and Okamoto (1969)). Two more recently introduced orderings are the more associated ordering of Schriever (1985) and the more totally positive of order 2 (TP_2) ordering of Kimeldorf and Sampson (1987). Hollander, Proschan and Sconing (1990) further discuss some of these orderings and introduce some related orderings. Another discussion of these orderings and the introduction of some new and related orderings are presented by Fang and Joe (1990). Other more specific orderings have been discussed by Shaked and Tong (1985), and Ruschendörf (1986). Results

[1] Research supported by NSA Grant No. R0909237. Reproduction in whole or part is permitted for any purpose of the United States Government.

concerning orderings can be found in Marshall and Olkin (1979), Tong (1980), and Kimeldorf and Sampson (1987), (1989). Other authors have considered partial orderings which correspond to more general notions of dependence, rather than just positive dependence. For example, Scarsini (1990) develops orderings involving the Lorenz curve of the likelihood ratio of a distribution with respect to the corresponding independence distribution. (See also Bromek and Kowalczyk (1990).) Joe (1985) develops an ordering for contingency tables based upon majorization.

Kimeldorf and Sampson (1987), proceeding in the spirit of Rényi (1959), provided a set of nine properties that an ordering on fixed marginal bivariate distributions should satisfy so that it may be called a *positive dependence ordering* (PDO). Yanagimoto (1990) developed an approach to generating orderings for dependence using a systematic definition, which is analogous to the approach for generalizing dependence concepts used in Yanagimoto (1972).

The purpose of this paper is to introduce another systematic and intuitive approach to generating orderings on bivariate distributions with fixed marginals. These orderings include the more concordant and the more TP_2 orderings, as well as all the orderings generated by the techniques of Yanagimoto (1990). We show that all of our orderings are partial orderings on bivariate distributions with fixed-marginals, and when appropriate are PDO's.

Throughout we assume that (X_1 , Y_1) is a random vector with c.d.f. $F(x , y)$ and (X_2 , Y_2) is a random vector with c.d.f. $G(x , y)$. We also assume that the x-marginals and the y-marginals are common, i.e., $F(x , \infty) = G(x , \infty)$ and $F(\infty , y) = G(\infty , y)$. When comparing two distributions, we equivalently write G is more ordered than F or (X_2 , Y_2) is more ordered than (X_1 , Y_1).

Let I_1 and J_1 be two real intervals. The notation $F(I_1 , J_1)$ is used instead of Prob$(X_1 \in I_1$ and $Y_1 \in J_1)$. Sometimes $F(I_1 , J_1)$ is further abbreviated $F(1 , 1)$. Also for any real sets S_1 and S_2, we write $S_1 < S_2$ if and only if $x_1 \in S_1$ and $x_2 \in S_2$ imply $x_1 < x_2$; similarly, we write $S_1 \leq S_2$ if the implication is $x_1 \leq x_2$.

2. Some Existing Orderings

The distribution G being more concordant than the distribution F is defined (Tchen (1980)) by

$$F(x , y) \leq G(x , y) , \tag{2.1}$$

for all x , y, and the notation $G \xrightarrow{c} F$ is employed. Kimeldorf and Sampson (1987) define G to be more TP_2 than F, if

$$F(1 , 1) F(2 , 2) G(1 , 2) G(2 , 1) \leq G(1 , 1) G(2 , 2) F(1 , 2) F(2 , 1) , \tag{2.2}$$

for all real intervals $I_1 < I_2$, $J_1 < J_2$; the notation used here is $G \xrightarrow{TP_2} F$.

Yanagimoto (1990) extends the more TP_2 ordering (2.2), utilizing the notions for positive dependence given in Yanagimoto (1972). To do so, Yanagimoto (1972), (1990) defined the following four families of products of intervals:

$$S(1) = \{(-\infty, x_1] \times (x_1, \infty) : -\infty < x_1 < \infty\} ,$$

$$S(2') = \{(x_1, x_2] \times (x_2, \infty) : -\infty < x_1 < x_2 < \infty\} ,$$

$$S(2'') = \{(-\infty, x_1] \times (x_1, x_2] : -\infty < x_1 < x_2 < \infty\} ,$$

and

$$S(3) = \{(x_1, x_2] \times (x_2, x_3] : -\infty < x_1 < x_2 < x_3 < \infty\} .$$

Then Yanagimoto (1990) defines $G(x, y)$ to have larger dependence than $F(x, y)$ in the sense of $P(i, j)$ if

$$F(1, 1)\, F(2, 2)\, G(1, 2)\, G(2, 1) \leq G(1, 1)\, G(2, 2)\, F(1, 2)\, F(2, 1) , \qquad (2.3)$$

for any $I_1 < I_2$ and $J_1 < J_2$ satisfying $I_1 \times I_2 \in S(i)$ and $J_1 \times J_2 \in S(j)$.

Thus there are 16 possible orders corresponding to
$(i, j) \in \{1, 2', 2'', 3\} \times \{1, 2', 2'', 3\}$.

Yanagimoto (1990) notes that $P(1, 1)$ is the more concordant ordering and $P(3, 3)$ is weaker than the more TP_2 ordering of Kimeldorf and Sampson (1987). He additionally notes a few other properties for and relationships among these 16 different orderings.

3. An Approach for Generating Orderings

In this section, we present another approach to generating orderings that allows us fairly easily to study their properties and interrelationships. The orderings are obviously motivated by Yanagimoto's (1990) approach, as well as, the "hieroglyphic-like" notation developed by Douglas, Fienberg, Lee, Sampson and Whitaker (1990) to study generalized odds-ratios for contingency tables.

We begin by introducing a general ordering approach that can be employed for any four classes of real intervals of the real line, denoted by I_1, I_2, J_1, J_2. Throughout we employ the abbreviation "A" to stand for "abut", where two disjoint intervals are said to abut if they share a common limit point. Also we use "NA" to stand for two disjoint intervals "not necessarily abuting," so that such intervals are not required to "abut," but may, in fact, do so.

Definition: G is said to be more dependent than F according to the ordering
$(I_1, I_2 ; J_1, J_2 ; NA, NA)$ if

$$F(I_1, J_1)\, F(I_2, J_2)\, G(I_1, J_2)\, G(I_2, J_1) \leq G(I_1, J_1)\, G(J_2, J_2)\, F(I_1, J_2)\, F(I_2, J_1) , \quad (3.1)$$

for all $I_1 \in I_1, I_2 \in I_2, J_1 \in J_1, J_2 \in J_2$, such that $I_1 < I_2$, and $J_1 < J_2$; that is, I_1 and I_2 and J_1 and J_2, respectively, may not abut. The notation

$$(X_2, Y_2) \xrightarrow{\;(I_1, I_2 ; J_1, J_2 ; NA, NA)\;} (X_1, Y_1) \text{ is used.}$$

The ordering $(I_1, I_2 ; J_1, J_2 ; NA, A)$ is defined as the preceding, with the additional constraint that limit $(J_1) \cap$ limit $(J_2) \neq \phi$, where for any interval K, limit (K) is the set of limit points of the interval K. In this definition, we note that the intervals J_1, J_2 are required to abut, but I_1, I_2 are not necessarily required to do this.

Similarly, define the orderings $(I_1, I_2; J_1, J_2; A, NA)$ and $(I_1, I_2; J_1, J_2; A, A)$, with the concomitant notation.

We now define the following three classes of real intervals

$$L = \{ L : L = (-\infty, a) \text{ or } (-\infty, a] : -\infty < a < \infty \} \qquad (3.2a)$$

$$R = \{ R : R = (a, \infty) \text{ or } [a, \infty) : -\infty < a < \infty \} \qquad (3.2b)$$

and

$$S = \{ S : S \text{ is any interval on the real line } \} ; \qquad (3.2c)$$

that is, L is the set of left intervals, R is the set of right intervals, and S is the set of strips, with the understanding $L \subset S$ and $R \subset S$.

We now combine the preceding three classes of intervals with the ordering definitions to provide ostensibly 64 possible orderings: $(L, R; L, R; NA, NA), \ldots,$ $(L, R; L, R; A, A)$, $(L, R; L, S; NA, NA), \ldots,$ $(L, R; S, R; NA, NA), \ldots,$ $(L, R; S, S; NA, NA), \ldots, (S, S; S, S; A, A)$.

It is direct to show that:

(i)   more concordant is $(L, R; L, R; A, A)$,

(ii)   more TP_2 is $(S, S; S, S; NA, NA)$,

(iii)   all $P(i, j)$ orderings can be written in this fashion.

With regard to (iii), $P(2'; 2')$, for example is, $(S, R; S, R; A, A)$, and $P(3, 2'')$ is $(S, S; L, S; A, A)$. The remaining translations can be made by noting the correspondences: $1 <=> L, R; 2 <=> S, R; 2'' <=> L, S;$ and $3 <=> S, S$. To deal with the issues of Yanagimoto's particular choice of open and closed intervals, in order to show equivalence to our orderings, one must use an argument like Lehmann (1966, p. 1138).

Note that

$$
\begin{array}{ccc}
 & (\cdot, \cdot; \cdot, \cdot; NA, A) & \\
\nearrow & & \searrow \\
(\cdot, \cdot; \cdot, \cdot; NA, NA) & & (\cdot, \cdot; \cdot, \cdot; A, A), \qquad (3.3) \\
\searrow & & \nearrow \\
 & (\cdot, \cdot; \cdot, \cdot; A, NA) &
\end{array}
$$

where any permissible combination of L, R, and S may be substituted for each pair "$\cdot, \cdot$". Additionally,

$$
\begin{array}{ccc}
 & (S, R; \cdots; \cdots) & \\
\nearrow & & \searrow \\
(S, S; \cdot, \cdot; \cdot, \cdot) & & (L, R; \cdot, \cdot; \cdot, \cdot), \qquad (3.4) \\
\searrow & & \nearrow \\
 & (L, S; \cdot, \cdot; \cdots) &
\end{array}
$$

where any permissible combination of L, R and S may be substituted for the first pair "$\cdot, \cdot$"

and any combination of A and NA for the second pair "· , · ". Similarly, one could hold fixed the second pair of classes and vary the first.

It is instructive to generate the graphical representations of these notions and view these orderings as comparisons of generalized odds-ratios for abuting and non-abuting sets. This is left for the reader's consideration.

4. PDO Properties for New Orderings

Kimeldorf and Sampson (1987) consider nine properties that an ordering, $G \rightarrow F$, for positive dependence might satisfy:

(P1) $G \rightarrow F \Rightarrow G(x, y) \geq F(x, y)$ for all x, y ;

(P2) $G \rightarrow F$ and $H \rightarrow G \Rightarrow H \rightarrow F$;

(P3) $F \rightarrow F$;

(P4) $F \rightarrow G$ and $G \rightarrow F \Rightarrow F = G$;

(P5) $F^+ \rightarrow F \rightarrow F^-$, where F^+ and F^- are, respectively, the corresponding upper and lower Fréchet bounds;

(P6) $(X_2, Y_2) \rightarrow (X_1, Y_1) \Rightarrow (h(X_2), Y_2) \rightarrow (h(X_1), Y_1)$, where h is a nondecreasing function;

(P7) $(X_2, Y_2) \rightarrow (X_1, Y_1) \Rightarrow (-X_1, Y_1) \rightarrow (-X_2, Y_2)$;

(P8) $(X_2, Y_2) \rightarrow (X_1, Y_1) \Rightarrow (Y_2, X_2) \rightarrow (Y_1, X_1)$; and

(P9) $G_n \xrightarrow{D} F_n$, $F_n \xrightarrow{D} F$, $G_n \xrightarrow{D} G \Rightarrow G \rightarrow F$, where $\rightarrow$ denotes convergence in distribu-

tion.

It is noted that in (P6), (P7), (P8) and (P9) the ordering $\rightarrow$ is applied to different fixed-marginal distributions.

Additionally, we note that (P2), (P3) and (P4) holding implies $\rightarrow$ is a partial ordering. Also (P8) basically requires the ordering to be symmetric in the random variables. The property (P1) says that the more concordant ordering is implied by any ordering $\rightarrow$.

If all nine properties are met by an ordering, Kimeldorf and Sampson (1987) term such an ordering a positive dependence ordering (PDO).

We now proceed to establish for each property which of our orderings satisfies it.

Property (P1). All of our orderings satisfy this. Because (3.3) and (3.4) hold, any ordering implies (L , R ; L , R ; A , A) which is the more concordant ordering.

Property (P2). All orderings satisfy this property. Kimeldorf and Sampson (1987, Lemma 4.1) showed that the TP_2 ordering (i.e., (S , S ; S , S ; NA , NA)) satisfies (P2). To do so, they showed transitivity for any fixed real intervals I_1 , I_2 , J_1 and J_2 such that $I_1 < I_2$, $J_1 < J_2$.

Since any of our orderings is defined like the TP_2 ordering only with some additional restrictions on the intervals, a close examination of their proof shows its applicability to this situation.

Property (P3). All orderings satisfy this trivially.

Property (P4). All orderings satisfy this from (P1) and the fact $\xrightarrow{c}$ is a PDO.

Property (P5). All orderings satisfy this and the proof is direct.

Property (P6). All orderings satisfy this property. This can be shown by breaking into cases: (i) h is continuous and strictly increasing; (ii) h is discontinuous and strictly increasing; (iii) h is not strictly increasing. In each of these cases, h maps left intervals, right intervals, and strips, respectively, into left intervals, right intervals, and strips, up to sets of marginal probability measure zero.

Property (P7). The orderings which satisfy this property are the 32 orderings (L ,R ; · , · ; · , ·) and (S , S ; · , · ; · , ·), where the first pair of "· , ·" can be L , R or S suitably combined and the second pair any combination of A and NA. To see this note that this property requires that if (P7) holds for I_1 , I_2 , it must also hold for $I_1^* = -I_2$, and $I_2^* = -I_1$, where $-K$ is the negative of the interval K. The only pairs satisfying this are (L , R) and (S , S).

Similarly, if we require the analogue of (P7) to hold in terms of $-Y_1$ and $-Y_2$, the only 32 orderings satisfying are of the form (· , · ; L , R ; · , ·) and (· , · ; S , S ; · , ·).

Property (P8). The orderings which satisfy this property are the 16 orderings (L , R ; L , R ; · , ·) , (L , S ; L , S ; · , ·) , (S , R ; S , R ; · , ·) and (S , S ; S , S ; · , ·). The reason for this is that if (P8) is to hold for I_1 , I_2 , J_1 , J_2 , it must also hold for J_1 , J_2 , I_1 , I_2 and thus in (3.1), $I_1 = J_1$ and $I_2 = J_2$.

Property (P9). All orderings can be shown to satisfy this property, by using the same argument that showed (P2).

Comments:

(i) The 8 orderings which are PDO's are (L , R ; L , R ; · , ·) and (S , S ; S , S ; · , ·).

(ii) All orderings are partial orderings, i.e., they satisfy (P2), (P3), (P4).

(iii) The 16 orderings which satisfy both (P7) and its analogue in terms of $-Y_1$ and $-Y_2$ are the 8 orderings in (i) above and the 8 orderings: (L , R ; S , S ; · , ·) and (S , S ; L , R ; · , ·).

5. Some Equivalences Among Our Orderings

As we indicated in Section 3, these 64 orderings are ostensibly different. For example, Yanagimota (1990) notes that (S , S ; S , S ; NA , NA) is "strictly stronger" than (S , S ; S , S ; A ,A), i.e., the TP$_2$ ordering is strictly stronger than P(3 , 3). Also we are not aware of any result which claims (L , R ; L , R ; A , A) is the same as (L , R ; L , R ; NA , NA). However, as we show in this section, both pairs of these noted orderings are, respectively, the same.

Theorem 5.1. (S , S ; S , S ; NA , NA) = (S , S ; S , S ; A , A), that is, Kimeldorf and Sampson's (1987) TP$_2$ ordering is the same as Yanagimoto's (1990) P(3 , 3).

Proof. We proceed by showing (S , S ; S , S ; A , A) $\Rightarrow$ (S , S ; S , S ; NA , NA) and use (3.3) to complete the proof.

Let $I_1 < I_2 < I_3$ be three abuting intervals on the x-axis and $J_1 < J_2 < J_3$ be three abuting intervals on the y-axis. Then G $\xrightarrow{\text{(S , S ; S , S ; A , A)}}$ F implies that the following four inequalities hold:

$$F(3 , 3) \, F(2 , 2) \, G(2 , 3) \, G(3 , 2) \le G(3 , 3) \, G(2 , 2) \, F(2 , 3) \, F(3 , 2) ; \qquad (5.1)$$

$$F(2 , 3) \, F(1 , 2) \, G(1 , 3) \, G(2 , 2) \le G(2 ,3) \, G(1 , 2) \, F(1 ,3) \, F(2 , 2) ; \qquad (5.2)$$

$$F(2 , 2) \, F(1 , 1) \, G(1 , 2) \, G(2 , 1) \le G(2 ,2) \, G(1 , 1) \, F(1 , 2) \, F(2 , 1) ; \qquad (5.3)$$

and

$$F(3 , 2) \, F(2 , 1) \, G(2 , 2) \, G(3 , 1) \le G(3 , 2) \, G(2 , 1) \, F(2 , 2) \, F(3 , 1) . \qquad (5.4)$$

Multiply together these four inequalities and cancel the common terms (assuming they are nonzero) to obtain

$$F(1 , 1) \, F(3 , 3) \, G(1 , 3) \, G(3 , 1) \le G(1 , 1) \, G(3 , 3) \, F(1 , 3) \, F(3 ,1) . \qquad (5.5)$$

If some common terms are zero, one needs to exercise slightly more care along the lines of Kimeldorf and Sampson (1987, Lemma 4.1).

Theorem 5.2. $(L , R ; L , R ; A , A) = (L ,R ; L , R ; NA , NA)$, that is Techen's (1980) more concordant ordering is the same as the ordering when left and right intervals are not necessarily abuting.

Proof. We show $(L , R ; L , R ; A , A) \overset{c}{\Rightarrow} (L , R ; L , R , NA , NA)$ and then use (3.3). It is well known that if $G \to F$, then

$$G(L_1 , L_2) \ge F(L_1 , L_2) ; \qquad (5.6)$$

$$G(R_1 , R_2) \ge F(R_1 , R_2) ; \qquad (5.7)$$

$$G(L_1 , R_2) \le F(L_1 , R_2) ; \qquad (5.8)$$

and

$$G(R_1 , L_2) \le F(R_1 , L_2) , \qquad (5.9)$$

where L_1 , L_2 are any left intervals and R_1 , R_2 are any right intervals. Let $I_1 < I_2$ and $J_1 < J_2$ be any not necessarily abuting intervals where I_1 and J_1 are left intervals and I_2 and J_2 are right intervals.

Apply (5.6) - (5.9), with $L_1 = I_1$, $L_2 = J_1$, $R_1 = I_2$ and $R_2 = J_2$. Multiplication of these four resultant inequalities yields that $G \xrightarrow{\,(L , R ; L , R ; NA , NA)\,} F$.

We note that Theorem 5.1 yields a slightly simpler way of checking whether or not G is more TP_2 than F, and Theorem 5.2 shows that not requiring abuting intervals in the more concordant ordering does not provide a stronger ordering. Clearly in both proofs the cases of equivalence also apply to concepts with A , NA and NA , A.

6. Discussion

The proofs of Theorem 5.1 and 5.2 require the specific structures and relationships of the sets defining these ordering concepts. We have not been able to show the same type of results hold for the other concepts, e.g., we do not know whether (L , S ; S , R ; A , A) and (L , S ; S , R ; NA , NA) are the same. Moreover, we have not been able to give counterexamples to these equivalences in these other cases. Clearly these issues warrant further investigation.

For $G \xrightarrow{c} F$, it is known (see Tchen (1980)) that this is equivalent to

E_G m(X , Y) $\geq E_F$ m(X , Y) for all suitable l-superadditive functions m. The generalizability of this type of concept to our other orderings is an open question. There is some general discussion of results of this nature in Kimeldorf and Sampson (1987) and Block, Chhetry, Fang and Sampson (1991).

As noted in Section 1, orderings have been used to explain and understand the parameter(s) in a variety of one-parameter families. For example, Ahmed, Langberg, Léon and Proschan (1979) and Tchen (1980) both examine a number of one-parameter families to see if these families are ordered according to the more concordant ordering. Fang and Joe (1990) examine questions of this nature for other related orderings. It would be of interest to examine a variety of one-parameter families in the context of the orderings considered herein.

Finally, we observe that it would be desirable to have conditions other than the definitions for checking whether these orderings hold in a particular case. That this may be a difficult task can be seen from the difficulties both Kimeldorf and Sampson (1987) and Yanagimoto (1990) have in developing a checkable density condition for the TP_2 ordering.

References

Ahmed, A., Langberg, N., Léon, R. and Proschan, F. (1979) 'Partial ordering of positive quadrant dependence with application', Technical Report, Florida State University.

Block, H. W., Chhetry, D., Fang, Z. and Sampson, A. R. (1991) 'Partial orders on permutations and dependence orderings on bivariate empirical distributions', to appear in *Ann. Statist.*

Bromek, T. and Kowalczyk (1990) 'A decision approach to ordering stochastic dependence', in, H. W. Block, A. R. Sampson, T. H. Savits (eds.), *Topics in Statistical Dependence*, Institute of Mathematical Statistics Lecture Notes - Monograph Series, Hayward, CA, (to appear).

Douglas, R., Fienberg, S. E., Lee. M-L T., Sampson, A. R. and Whitaker, L. R. (1990) 'Positive dependence concepts for ordinal contingency tables', in, H. W. Block, A. R. Sampson, T. H. Savits (eds.), *Topics in Statistical Dependence*, Institute of Mathematical Statistics Lecture Notes - - Monograph Series, Hayward, CA, (to appear).

Fang, Z. and Joe, H. (1990) 'Further developments on dependence orderings for continuous bivariate distributions', Department of Statistics, University of British Columbia, Technical Report 87.

Hollander, M., Proschan, F. and Sconing, J. (1990) 'Information, censoring, and dependence', in, H. W. Block, A. R. Sampson, T. H. Savits (eds.), *Topics in Statisical Dependence*, Institute of Mathematical Statistics Lecture Notes - Monograph Series, Hayward, CA (to appear).

Joe, H. (1985) 'An ordering of dependence for contingency tables', *Linear Algebra and Its Applications* **70**, 89-103.

Kimeldorf, G. and Sampson, A. R. (1987) 'Positive dependence orderings', *Ann. Inst. Statist. Math* **39**, 113-128.

Kimeldorf, G. and Sampson, A. R. (1989) 'A framework for positive dependence', *Ann. Inst. Statist. Math* **41**, 31-45.

Lehmann, E. (1966) 'Some concepts of dependence', *Ann. Math. Statist.* **37**, 1137-1153.

Marshall, A. W. and Olkin, I (1979) *Inequalities: Theory of Majorization and Its Application*, Academic Press, New York.

Metry, M. H. and Sampson, A. R. (1988). 'Positive dependence concepts for multivariate empirical rank distributions', University of Pittsburgh, Department of Mathematics and Statistics, Technical Report 88-09.

Rényi, A. (1959) 'On measures of dependence', *Acta Math. Acad. Sci. Hungar* **10**, 441-451.

Scarsini, M. (1990) 'An ordering of dependence', in, H. W. Block, A. R. Sampson, T. H. Savits (eds.), *Topics in Statistical Dependence*, Institute of Mathematical Statisitics Lecture Notes - Monograph Series, Hayward, CA, (to appear).

Rüschendorf, L. (1986) 'Monotonicity and unbiasedness of test via a.s. constructions', *Statistics* **17**, 221-230.

Schriever, B. F. (1985) *Order Dependence*, Ph.D. Dissertation, Free University of Amsterdam.

Schriever, B. F. (1987a) 'Monotonicity of rank statistics in some nonparametric testing problems', *Statistica Neerlandica* **41**, 99-109.

Schriever, B. F. (1987b) 'An ordering for positive dependence', *Ann. Statist.* **15**, 1208-1214.

Shaked, M. and Tong, Y. L. (1985) 'Some partial orderings of exchangeable random variables by positive dependence', *J. Multivar. Anal.* **17**, 333-349.

Tchen, A. H. T. (1980) 'Inequalities for distribution with given marginals', *Ann. Prob.* **8**, 814-827.

Tong, Y. L. (1980) *Probability Inequalities in Multivariate Distributions*, Academic Press, New York.

Yanagimoto, T. and Okamoto, M. (1969) 'Partial orderings of permutations and monotonicity of a rank correlation statistic', *Ann. Inst. Statist. Math.* **21**, 489-506.

Yanagimoto, T. (1972) 'Families of positively dependent random variables', *Ann. Inst. Statist. Math.* **24**, 559-573.

Yanagimoto, T. (1990) 'Dependence ordering in statistical models and other notions', in, H. W. Block, A. R. Sampson, T. H. Savits (eds.), *Topics in Statistical Dependence*, Institute of Mathematical Statistics Lecture Notes - Monograph Series, Hayward, CA, (to appear).

INDECOMPOSABLE MARGINAL PROBLEMS

HANS G. KELLERER
Department of Mathematics
University of Munich
Theresienstrasse 39
8000 München 2
Germany

ABSTRACT. It is shown that duality theorems for marginal problems that hold in the decomposable case for all bounded Borel measurable functions extend to the indecomposable case, if the function under consideration is upper semicontinuous, but may fail, if it is lower semicontinuous.

Introduction

Let be given

(1)   a finite index set T and a family $\mathfrak{T}$ of nonempty subsets U of T,

(2)   Hausdorff spaces X_i, $i \in T$, with product spaces $X_U = \prod_{i \in U} X_i$,

(3)   finite Borel measures μ_U on X_U for $U \in \mathfrak{T}$.

Then the classical marginal problem concerns the existence of a finite Borel measure μ on the whole product $X = X_T$ having marginals

$$\pi_U(\mu) = \mu_U \qquad \text{for all } U \in \mathfrak{T},$$

where π_U denotes the canonical projection of X onto X_U.

Obviously it is no real restriction to assume henceforth

(4) $\mathfrak{T} \neq \emptyset$ and $\bigcup \mathfrak{T} = T$.

If, in addition, the sets in $\mathfrak{T}$ are pairwise disjoint and the measures are normalized by $\mu_U(X_U) = 1$, the problem admits a trivial solution, given by the product measure $\mu = \bigotimes_{U \in \mathfrak{T}} \mu_U$ (provided the spaces X_i have a countable base). Otherwise the measures μ_U, $U \in \mathfrak{T}$, have at least to satisfy the consistency condition

(5) $\pi_V(\mu_{U_1}) = \pi_V(\mu_{U_2})$ for $U_k \in \mathfrak{T}$ and $V = U_1 \cap U_2$,

139

where for $U \supset V$ the projection of X_U onto X_V is simply denoted be π_V and for $V = \emptyset$ equation (5) has to be understood as

$$\mu_{U_1}(X_{U_1}) = \mu_{U_2}(X_{U_2}) \; .$$

Now it turns out that even in the simplest nontrivial case

$$T = \{1,2,3\} \qquad \text{and} \qquad \mathfrak{T} = \{ \{1,2\}, \{2,3\} \}$$

condition (5) does not guarantee the existence of a solution μ. A counter-example in [5] shows that this in fact can occur for metrizable spaces X_i with a countable base. Therefore this study will be carried out within the class of finite Radon measures, denoted by $M(Y)$ for any Hausdorff space Y and consisting of all finite measures on the Borel σ-algebra $\mathfrak{B}(Y)$ of Y satisfying the condition

$$\mu(B) = \sup \{\mu(K): K \subset B \text{ compact}\} \qquad \text{for all } B \in \mathfrak{B}(Y) \; .$$

Then it turns out that, given only T and $\mathfrak{T}$, the question whether condition (5) is sufficient for the existence of a measure $\mu \in M(X)$ with given marginals $\mu_U \in M(X_U)$, $U \in \mathfrak{T}$, is of a purely combinatorial nature, the answer being independent of the spaces X_i and the measures μ_U. The crucial property of the "hypergraph" $\mathfrak{T}$ is its "decomposability", defined as follows: there exists an arrangement $\mathfrak{T} = \{U_1, \ldots, U_n\}$ such that

$$(\bigcup_{1 \le m} U_l) \cap U_m \in \bigcup_{1 \le m} \mathfrak{P}(U_l) \qquad \text{for } 1 \le m \le n \; ,$$

where $\mathfrak{P}(A)$ denotes the power set of any set A. This result dates back to independent work by Vorobev [12] (in the case of finite spaces) and the author [4] (in the case $X_i = \mathbb{R}$). Later the subject was taken up by Lauritzen et al. [7] and has gained importance in the context of expert systems (see e.g. [8]).

Much more involved is the indecomposable case, considered for the first time by Dall'Aglio [1] in the special situation

$$|T| = 3 \qquad \text{and} \qquad \mathfrak{T} = \{U \subset T: |U| = 2\} \; .$$

While he obtained necessary conditions only, a complete criterion was given by the author for the real case in [4] and for the general case in [5] (see also [10], [9], [3]). Confined to completely regular spaces X_i it runs as follows: a solution of the marginal problem exists if and only if

$$(6) \qquad \sum_{U \in \mathfrak{T}} \int_{X_U} g_U \, d\mu_U \ge 0 \qquad \text{for all } g_U \in \mathscr{C}(X_U) \text{ with } \sum_{U \in \mathfrak{T}} g_U \circ \pi_U \ge 0,$$

where $\mathscr{C}(Y)$ denotes the space of bounded continuous functions on any Hausdorff space Y.

The sufficiency of condition (6), whose necessity is straightforward, can also be derived from [6] by maximizing the total mass $\mu(X)$ among all measures μ, whose marginals do not exceed the given measures μ_U, $U \in \mathfrak{X}$. In this way the existence problem becomes a special case of a measure theoretic version of linear programming, defined as follows: given some function $h \in \mathscr{C}(X)$, consider the dual pair of infinite linear programs

(7) $\qquad \begin{cases} \text{maximize } \int\limits_X h d\mu \text{ under the constraint} \\ \pi_U(\mu) \preceq \mu_U \qquad \text{for } U \in \mathfrak{X}, \end{cases}$

(8) $\qquad \begin{cases} \text{minimize } \sum\limits_{U \in \mathfrak{X}} \int\limits_{X_U} g_U d\mu_U \text{ under the constraint} \\ 0 \leq g_U \in \mathscr{C}(X_U) \qquad \text{and} \qquad \sum\limits_{U \in \mathfrak{X}} g_U \circ \pi_U \succeq h \end{cases}$

and investigate the duality equation sup (7) = inf (8) .

It is a consequence of the Hahn-Banach theorem that this equation indeed holds true (provided the spaces X_i are completely regular), and it is proved in [6] that it remains valid, if h is any bounded Borel measurable functions and $\mathscr{C}(X_U)$ in (8) is replaced by the space $\mathscr{L}(\mu_U)$ of μ_U-integrable functions (provided the spaces X_i are metrizable). Moreover, it can be deduced that this continues to hold, if the constraint in (7) is sharpened by requiring equality of $\pi_U(\mu)$ and μ_U and the constraint in (8) is weakened by dropping the condition $g_U \succeq 0$ - this time, however, only in the decomposable case, the marginals being consistent.

It is the aim of this paper to study the duality equation in the indecomposable case and with marginals prescribed exactly, thus combining the existence and the optimization problem. Two main results will be established:

(+) duality holds for upper semicontinuous functions h, if only the spaces X_i are completely regular,

(-) duality may fail for lower semicontinuous functions h, even if the spaces X_i are compact and metrizable.

Incidentally, the last result disproves a statement in [11, p. 269].

The main result

Since the assumptions on the spaces X_i and the measures μ_U varied in the course of the introduction, they will be fixed now for the rest of the paper: in addition to (1)-(4) henceforth

(9) X_i is completely regular for $i \in T$,

(10) $\mu_U \in M(X_U)$ for $U \in \mathfrak{T}$.

Then the following functionals are of central interest:

Definition 1 *Let* h *be a function from* X *to the extended real line* $\tilde{\mathbb{R}}$; *then*

$$S(h) = \sup \{ \int_X^* h d\mu : \mu \in M(X) \ \text{with} \ \pi_U(\mu) = \mu_U \ \text{for} \ U \in \mathfrak{T} \},$$

$$I(h) = \inf \{ \sum_{U \in \mathfrak{T}} \int_{X_U} g_U d\mu_U \ g_U \in \mathscr{L}(\mu_U) \ \text{with} \ \sum_{U \in \mathfrak{T}} g_U \circ \pi_U \geq h \}.$$

Here, three remarks are in order:

(a) if the relevant set is empty, as usual $S(h) = -\infty$ resp. $I(h) = +\infty$,

(b) since h need not be contained in $\mathscr{L}(\mu)$, the upper integral

$$\int_X^* h d\mu = \inf \{ \int_X f d\mu : f \in \mathscr{L}(\mu) \ \text{with} \ f \geq h \}$$

has to be used (and may attain an infinite value),

(c) to ensure the sum $\sum_{U \in \mathfrak{T}} g_U \circ \pi_U$ to be well-defined everywhere, for the functions g_U the value $+\infty$ is allowed (on null sets) but the value $-\infty$ is always excluded.

One half of the duality equation is trivial: if $S(h) > -\infty$ and $I(h) < +\infty$, there exist $\mu \in M(X)$ and $g_U \in \mathscr{L}(\mu_U)$ according to Definition 1 and every choice yields

$$\int_X^* h d\mu \leq \int_X (\sum_{U \in \mathfrak{T}} g_U \circ \pi_U) d\mu$$

$$= \sum_{U \in \mathfrak{T}} \int_{X_U} g_U \, d\pi_U(\mu)$$

$$= \sum_{U \in \mathfrak{T}} \int_{X_U} g_U \, d\mu_U .$$

Moreover, in establishing the inverse inequality $S(h) \geq I(h)$ the assumption $I(h) > -\infty$ is no restriction.

Now it turns out that for a large class of functions h this inequality can in fact be proved in a sharper form. To this end the functional I has to be replaced by its topological version:

Definition 2 *Let* h: $X \longrightarrow \tilde{\mathbb{R}}$ *be arbitrary; then:*

$$I_c(h) = \inf \{ \sum_{U \in \mathfrak{T}} \int_{X_U} g_U \, d\mu_U : g_U \in \mathscr{C}(X_U) \ \text{with} \ \sum_{U \in \mathfrak{T}} g_U \circ \pi_U \geq h \}.$$

Due to $\mathscr{C}(X_U) \subset \mathscr{L}(\mu_U)$ clearly $I_c(h) \geq I(h)$.

Now the proof of the inequality $S(h) \geq I_c(h)$ splits into two steps. The first one is pure functional analysis:

Lemma 1 *Let* $L_U: \mathscr{C}(X_U) \longrightarrow \mathbb{R}$ *be a positive linear functional for*

$U \in \mathfrak{T}$ *and let* $h \in \mathscr{C}(X)$ *satisfy the condition*

$$\eta = \inf \{ \sum_{U \in \mathfrak{T}} L_U(g_U) : g_U \in \mathscr{C}(X_U) \text{ with } \sum_{U \in \mathfrak{T}} g_U \circ \pi_U \geq h \} > -\infty.$$

Then there exists a positive linear functional $L: \mathscr{C}(X) \longrightarrow \mathbb{R}$ *such that*

(11) $\qquad L(g_U \circ \pi_U) = L_U(g_U) \qquad$ *for* $g_U \in \mathscr{C}(X_U)$, $U \in \mathfrak{T}$,

(12) $\qquad L(h) = \eta$.

Proof 1. For $f \in \mathscr{C}(X)$ define

$$N(f) = \inf \{ \sum_{U \in \mathfrak{T}} L_U(g_U) - \gamma\eta :$$

$$g_U \in \mathscr{C}(X_U) \text{ and } \gamma \geq 0 \text{ with } \sum_{U \in \mathfrak{T}} g_U \circ \pi_U - \gamma h \geq f \}.$$

This functional has the following properties:

(13) $\qquad N(f) < +\infty \qquad$ for all $f \in \mathscr{C}(X)$,

because the relevant set is nonempty;

(14) $\qquad N(f) \leq 0 \qquad$ for $f \in \mathscr{C}(X)$ with $f \leq 0$,

because in this case $g_U = 0$ and $\gamma = 0$ are admissible;

(15) $\qquad N(\alpha f) = \alpha N(f) \qquad$ for $\alpha > 0$ and $f \in \mathscr{C}(X)$;

(16) $\qquad N(f_1 + f_2) \leq N(f_1) + N(f_2) \qquad$ for $f_k \in \mathscr{C}(X)$;

(17) $\qquad N(h) \geq \eta$,

because for $g_U \in \mathscr{C}(X_U)$ and $\gamma \geq 0$ with

$$\sum_{U \in \mathfrak{T}} g_U \circ \pi_U - \gamma h \geq h$$

the inequality

$$\sum_{U \in \mathfrak{T}} (\frac{1}{1+\gamma} g_U) \circ \pi_U \geq h$$

holds and implies

$$\sum_{U \in \mathfrak{T}} \frac{1}{1+\gamma} L_U(g_U) \geq \eta$$

or equivalently

$$\sum_{U \in \mathfrak{T}} L_U(g_U) - \gamma\eta \geq \eta;$$

(18) $\qquad N(f) > -\infty \qquad$ for all $f \in \mathscr{C}(X)$,

because due to (16) and (17)

$$N(f) + N(h - f) \geq N(h) > -\infty.$$

2. By part 1 N is a finite-valued positively homogeneous and sub-additive functional, hence the Hahn-Banach theorem applies and yields a linear functional L such that

$$L(f) \le N(f) \qquad \text{for all } f \in \mathscr{C}(X),$$

which due to (14) is also positive. In addition,

$$L(\sum_{U \in \mathfrak{T}} g_U \circ \pi_U - \gamma h) \le N(\sum_{U \in \mathfrak{T}} g_U \circ \pi_U - \gamma h)$$
$$\le \sum_{U \in \mathfrak{T}} L_U(g_U) - \gamma \eta \text{ for } g_U \in \mathscr{C}(X_U) \text{ and } \gamma \ge 0$$

implies

$$L(g_U \circ \pi_U) \le L_U(g_U) \qquad \text{for all } g_U \in \mathscr{C}(X_U), U \in \mathfrak{T},$$

which yields (11) immediately, and

$$L(-\gamma h) \le -\gamma \eta \qquad \text{for all } \gamma \ge 0,$$

which yields (12), because the inequality $L(h) \le \eta$ is obvious. $\square$

At the next step topological measure theory enters:

Lemma 2   Let $L: \mathscr{C}(X) \longrightarrow \mathbb{R}$ be a positive linear functional satisfying

$$L(g_U \circ \pi_U) = \int_{X_U} g_U \, d\mu_U \qquad \text{for } g_U \in \mathscr{C}(X_U), U \in \mathfrak{T}.$$

Then there exists a measure $\mu \in M(X)$ such that

$$L(f) = \int_X f \, d\mu \qquad \text{for all } f \in \mathscr{C}(X).$$

Proof In view of the Daniell extension procedure it suffices to show that the functional L is "tight", i.e. for arbitrary $\epsilon > 0$ there exists a compact subset K of X such that

$$L(f) < \epsilon \qquad \text{for all } f \in \mathscr{C}(X) \text{ with } 0 \le f \le 1_{X \setminus K}.$$

To this end denote by μ_i, $i \in T$, the one-dimensional marginals associated with μ_U, $U \in \mathfrak{T}$, which are uniquely determined and belong to $M(X_i)$ in view of (9), (10) and $\bigcup \mathfrak{T} = T$. Next fix $\delta > 0$ and

$$K = \prod_{i \in T} K_i \qquad \text{with } K_i \subset X_i \text{ compact.}$$

Then for a function $f \in \mathscr{C}(X)$ with $0 \le f \le 1_{X \setminus K}$ the set $G = \{f < \delta\}$ is an open neighbourhood of K, hence by a theorem of Wallace (see e.g. [2, p. 140])

$$K \subset \prod_{i \in T} G_i \subset G \qquad \text{with } G_i \subset X_i \text{ open.}$$

Moreover, in view of (9) there are functions $g_i \in \mathscr{C}(X_i)$ with $0 \le g_i \le 1$ such that

$$g_i = 0 \text{ on } K_i \qquad \text{and} \qquad g_i = 1 \text{ on } X_i \setminus G_i$$

(see e.g. [2, p. 124]). Together this implies

$$f \leq \delta 1_G + \sum_{U \in \mathfrak{X}} 1_{X_i \setminus G_i} \circ \pi_i$$

$$\leq \delta + \sum_{U \in \mathfrak{X}} g_i \circ \pi_i.$$

Applying L this yields

$$L(f) \leq L(\delta) + \sum_{i \in T} \int_{X_i} g_i d\mu_i$$

$$\leq \delta L(1) + \sum_{i \in T} \mu_i(X_i \setminus K_i).$$

Since the last sum can be decreased below ε by suitable choice of δ and K_i, the proof is finished. $\square$

Now the first main result can be established:

Proposition 1   Let h: $X \longrightarrow \widetilde{\mathbb{R}}$ *be upper semicontinuous with*

$$\sup_{x \in X} h(x) < + \infty \qquad and \qquad I_c(h) > -\infty.$$

Then there exists a measure $\mu \in M(X)$ *such that*

$$\pi_U(\mu) = \mu_U \qquad for \ U \in \mathfrak{X},$$

$$\int_X h d\mu = I_c(h).$$

Proof   Since X is completely regular, the set

$$\mathcal{H} = \{f \in \mathscr{C}(X): f \geq h\}$$

decreases to h. If h is replaced by any function $f \in \mathcal{H}$, the positive linear functionals

$$L_U(g_U) = \int_{X_U} g_U \, d\mu_U \qquad for \ g_U \in \mathscr{C}(X_U), \ U \in \mathfrak{X}$$

satisfy the condition of Lemma 1. Combined with Lemma 2 this yields measures $\mu_f \in M(X)$ with the given marginals such that

$$\int_X f d\mu_f \geq I_c(h) \qquad for \ f \in \mathcal{H}.$$

Therefore, the sets

$$M_f = \{\mu \in M(X): \pi_U(\mu) = \mu_U \ for \ U \in \mathfrak{X} \ and \ \int_X f d\mu \geq I_c(h)\}$$

are nonempty and clearly satisfy

$$M_{f_1} \cap M_{f_2} \supset M_{f_1 \wedge f_2} \qquad for \ f_k \in \mathcal{H}.$$

Moreover, they are closed with respect to the weak (or narrow) topology and included in the set

$$M = \{\mu \in M(X): \pi_i(\mu) = \mu_i \ for \ i \in T\},$$

where again μ_i, $i \in I$, are the one-dimensional marginals associated with μ_U, $U \in \mathfrak{X}$. Since M is compact (see [5, p. 402]), this implies

$$M_0 = \bigcap_{f \in \mathcal{H}} M_f \neq \emptyset.$$

But any $\mu \in M_0$ is τ-continuous, thus

$$\int_X h \, d\mu = \inf_{f \in \mathcal{H}} \int_X f \, d\mu \geq I_c(h),$$

and this concludes the proof, the inverse inequality being obvious. $\quad\square$

It should be mentioned that Proposition 1 in particular solves the existence problem. Indeed, if the necessary condition (6) is satisfied, clearly $I_c(0) \geq 0 \; (> -\infty)$, and thus there exists a measure $\mu \in M(X)$ with marginals μ_U, $U \in \mathfrak{T}$ (and $\int_X 0 \, d\mu = I_c(0)$).

Extensions

It is seen by the trivial example $\mathfrak{T} = \{T\}$ that an equation $S(h) = I_c(h)$ cannot be expected, unless h is upper semicontinuous. Therefore, in more general situations the functional I_c has to be replaced by I. As will be seen, however, the only extension thus made possible concerns the boundedness condition. More precisely:

Proposition 2   *Let* $h: X \longrightarrow \tilde{\mathbb{R}}$ *be upper semicontinuous with*

$$-\infty < I(h) < +\infty.$$

Then there exists a measure $\mu \in M(X)$ *such that*

$$\pi_U(\mu) = \mu_U \qquad \textit{for } U \in \mathfrak{T},$$

$$\int_X h \, d\mu = I(h).$$

Proof   According to the assumption there are functions $g_U \in \mathcal{L}(\mu_U)$ such that

$$h \leq \sum_{U \in \mathfrak{T}} g_U \circ \pi_U.$$

Since the measures μ_U are regular, the functions g_U can be approximated from above by lower semicontinuous functions $g'_U \in \mathcal{L}(\mu_U)$. Then

$$g' = \sum_{U \in \mathfrak{T}} g'_U \circ \pi_U$$

is itself lower semicontinuous and according to convention (c) following Definition 1 does not attain the value $-\infty$. Therefore the function

$$h'(x) = \begin{cases} h(x) - g'(x) & \text{if } h(x) < +\infty \\ 0 & \text{otherwise} \end{cases}$$

is well-defined, bounded above by 0, and upper semicontinuous as follows from the equation

$$h' = \inf_{n \in \mathbb{N}} ((h \wedge n) - (g' \wedge n)).$$

Now functions $g_u \in \mathscr{C}(X_u)$ with

$$\sum_{u \in \mathfrak{X}} g_u \circ \pi_u \geq h'$$

satisfy everywhere the inequality

$$\sum_{u \in \mathfrak{X}} g_u \circ \pi_u + \sum_{u \in \mathfrak{X}} g'_u \circ \pi_u \geq h.$$

With the abbreviation

$$\gamma' = \sum_{u \in \mathfrak{X}} \int_{X_u} g'_u d\mu_u$$

this yields the lower bound

$$I_c(h') \geq I(h) - \gamma' > -\infty.$$

Therefore, by Proposition 1 there exists $\mu \in M(X)$ with the given marginals and

$$\int_X h' d\mu = I_c(h').$$

Then $g' \in \mathscr{L}(\mu)$, hence $\{g' = +\infty\}$ is a null set modulo μ and in view of $h \leq g'$ this holds for $\{h = +\infty\}$ as well. Therefore $h = h' + g'$ modulo μ, thus

$$\int_X h d\mu = I_c(h') + \gamma' \geq I(h),$$

and this concludes the proof, the inverse inequality being obvious. $\square$

It remains to show by a counterexample that Proposition 2 is in a way best possible. Indeed, already in the simplest indecomposable case there are compact metrizable spaces X_i and a bounded lower semi-continuous function h with a nontrivial duality gap, i.e.

$$-\infty < S(h) < I(h) < +\infty.$$

In fact an indicator function may be chosen for h:

Example *Let* T *and* $\mathfrak{X}$ *be given by*

$$|T| = 3 \qquad and \qquad \mathfrak{X} = \{U \subset T: |U| = 2\}.$$

For $i \in T$ *let* X_i *be the unit interval and for* $U \in \mathfrak{X}$ *let* μ_U *be the equidistribution on the subset* $\{\sum_{i \in U} x_i \leq 1\}$ *of* X_U. *Then the indicator function* h *of the open subset* $G = \{\sum_{i \in T} x_i \neq 1\}$ *of* X *satisfies*

$$S(h) = 0 \qquad and \qquad I(h) = 1.$$

Proof 1. If μ is the equidistribution on the closed set $F = X \setminus G$, it is easily checked that μ has the marginals μ_U, $U \in \mathfrak{X}$. If $\mu' \in M(X)$ is any other measure with this property, then

$$\int_X (\sum_{i \in T} x_i - 1)^2 \, d\mu' = \int_X (\sum_{i \in T} x_i - 1)^2 \, d\mu = 0,$$

because these integrals depend on the measures only via their two-dimensional marginals. Therefore μ' is supported by F as well and $\mu' = \mu$ is an immediate consequence. Since $\mu(G) = 0$, this proves the equation $S(h) = 0$.

2. Next let $g_U \in \mathscr{L}(\mu_U)$ satisfy

(19) $\sum_{U \in \mathfrak{X}} g_U \circ \pi_U \geq h = 1_G.$

Partition each space X_i into n intervals of equal length and denote by g_U^n the conditional expectations of g_U with respect to the corresponding partition of X_U and two-dimensional Lebesgue measure. Since F is a null set with respect to three-dimensional Lebesgue measure, an integration of (19) by this measure yields

$$\sum_{U \in \mathfrak{X}} g_U^n \circ \pi_U \geq 1.$$

Another integration by the measure μ from part 1 transforms this into

(20) $\sum_{U \in \mathfrak{X}} \int_{X_U} g_U^n \, d\mu_U \geq 1.$

Now a passage from g_U^n to g_U affects the integral only via the contribution over the union D_U^n of the n elements of the nth partition of X_U along the diagonal $\{\sum_{i \in U} x_i = 1\}$. More precisely, it is easily verified that

$$| \int_{X_U} g_U^n \, d\mu_U - \int_{X_U} g_U \, d\mu_U | \leq \int_{D_U^n} |g_U| \, d\mu_U.$$

Since $\mu_U(D_U^n) \to 0$ for $n \to \infty$, it follows from (20) that

$$\sum_{U \in \mathfrak{X}} \int_{X_U} g_U \, d\mu_U \geq 1.$$

This proves the inequality $I(h) \geq 1$, while the converse is trivial. $\square$

In conclusion, two remarks are in order. First, Propositions 1 and 2 not only establish the duality equation $S(h) = I(h)$ but in addition prove the supremum to be a maximum. The analogous question concerning the infimum, answered in the affirmative in [6] for the decomposable case, is left open, however, in the present context. The crucial point lies in the fact that, while the functional S is continuous with respect to decreasing sequences, the functional I fails to be continuous with respect to increasing sequences.

Finally, it has to be mentioned that Propositions 1 and 2 can be extended to arbitrary index sets T as long as $\mathfrak{X} \subset \mathfrak{P}(T)$ is countable. In this case the majorization of h has to be carried out by partial sums $\sum_{U \in \mathfrak{U}} g_U \circ \pi_U$ with $\mathfrak{U} \subset \mathfrak{T}$ finite. A further generalization fails, because a projective system $\mu_U \in M(X_U)$, $U \subset T$ finite, need not have a projective limit $\mu \in M(X)$ for uncountable T.

References

1. Dall'Aglio, G. (1959) 'Sulla compatibilità delle funzioni di ripartizione doppia', Rend. Mat. 18, 385-413.
2. Engelking, R. (1989) General Topology, Heldermann Verlag, Berlin.
3. Gaffke, N. and Rüschendorf, L. (1984) 'On the existence of probability measures with given marginals', Statist. Decisions 2, 163-174.
4. Kellerer, H.G. (1964) 'Verteilungsfunktionen mit gegebenen Marginalverteilungen', Z. Wahrsch. Verw. Gebiete 3, 247-270.
5. —— (1984) 'Duality theorems for marginal problems', Z. Wahrsch. Verw. Gebiete 67, 399-432.
6. —— (1988) 'Measure theoretic versions of linear programming', Math. Z. 198, 367-400.
7. Lauritzen, S.L., Speed, T.P., Vijayan, K. (1984) 'Decomposable graphs and hypergraphs', J. Austr. Math. Soc. (Ser. A) 36, 12-29.
8. Lauritzen, S.L. and Spiegelhalter, D.J. (1988) 'Local computations with probabilities on graphical structures and their application to expert systems', J. Roy. Statist. Soc. (Ser. B) 50, 157-224.
9. Lembcke, J. (1977) 'Reguläre Maße mit einer gegebenen Familie von Bildmaßen', Sitzungsber. Bayer. Akad. Wiss. 1976, 61-115.
10. Maharam, D. (1971) 'Consistent extensions of linear functionals and of probability measures', Proc. 6th Berkeley Sympos. Math. Statist. Prob. 2, 127-147.
11. Rüschendorf, L. (1984) 'On the minimum discrimination information theorem', Statist. Decisions (Suppl.) 1, 263-283.
12. Vorobev, N.N. (1962) 'Consistent families of measures and their extensions', Theory Probab. Appl. 7, 147-163.

FRÉCHET-BOUNDS AND THEIR APPLICATIONS

Ludger Rüschendorf
Inst. für Math. Statistik
Einsteinstr. 62, D-4400 Münster

Summary. This paper gives a review of Fréchet-bounds and their applications. In section two an approach to the marginal problem and Fréchet-bounds based on duality theory resp. the Hahn-Banach theorem is discussed. Main applications concern the Strassen representation theorem for stochastic orders, the sharpness of the classical Fréchet-bounds, the representation of minimal metrics, couplings of distributions, the Monge-Kantorovic-problem, the construction of random variables with maximum (resp. minimum) sum and variances of the sum, maximally dependent random variables and others. For multivariate marginal systems there is a useful reduction principle and there are some bounds for simple systems, which yield a characterization of the marginal problem for a system of two dimensional marginals in a three-fold product space. In section three we discuss some generalizations of the Young-inequality, which are useful for solving the dual problems of the Fréchet-bounds. A basic notion in this connection is the notion of c-convex functions. As an application one can give a nice characterization of solutions of certain transportation problems. We give a probabilistic proof of some generalizations of the Young- and the Oppenheim-inequality. In section four we discuss some statistical applications and problems. The Huzurbazar conjecture on marginal sufficiency, the problem of the optimal combination of marginal tests and the question of estimation theory in marginal models is considered.

1.   Introduction

The marginal model is formally defined as follows. Let $E = E_1 \times \ldots \times E_n$, $\mathfrak{A} = \mathfrak{A}_1 \otimes \ldots \otimes \mathfrak{A}_n$ be a finite product of measure spaces. Let $\mathfrak{E} \subset \mathfrak{P}(\{1,\ldots,n\})$, the system of all subsets of $\{1,\ldots,n\}$ with $\bigcup_{J \in \mathfrak{E}} J = \{1,\ldots,n\}$ and let for $J \in \mathfrak{E}$, $P_J \in M(\prod_{j \in J} E_j)$ be a consistent system of probability measures on $\pi_J(E) = \prod_{j \in J} E_j =: E_J$, where π_J denotes the J-projection from E to E_J. Define the marginal model $M_{\mathfrak{E}}$:

(1) $$M_{\mathbf{C}} = M(P_J, J \in \mathbf{C}) = \{P \in M\ (E, \mathbf{A});\ P^{\Pi_J} = P_J,\ J \in \mathbf{C}\}$$

to be the set of all probability measures on E with marginals $P_J = P^{\Pi_J}$ of the J-components, $J \in \mathbf{C}$.

There are some different type of problems of interest in marginal models and related to Frechet-bounds. The _marginal problem_ is the question, whether $M_{\mathbf{C}} \neq \emptyset$. It was shown by Vorobev (1962), Kellerer (1964), that the property "consistency of $(P_J)_{J \in \mathbf{C}}$ implies $M_{\mathbf{C}} \neq \emptyset$" is a purely combinatorial (graphtheoretic) property and is equivalent to the nonexistence of "cycles" in $\mathbf{C}$. Systems $\mathbf{C}$ with this property are called _decomposable_ resp. _simplicial complex_ (in [100]). Some related existence problems are investigated in [46], [26], [39], [35]. For non-regular systems the marginal problem is generally not easy to decide except in cases, where explicit constructions are known (cf. [14], [84]). Generally, $M_{\mathbf{C}}$ is a convex set of probability measures, which in a topological situation with tight P_J is also compact. From a theorem of Douglas (1964), $P \in M_{\mathbf{C}}$ is an extreme point iff $F = \{\sum_{J \in \mathbf{C}} f_J \circ \pi_J;\ f_J \in \mathbf{B}^1(P_J)\}$ is dense in $\mathbf{B}^1(P)$. For simple marginals more information is known if n = 2. Generally, $M_{\mathbf{C}}$ can be empty, can be a small (even one-point) set or can be a large set of distributions.

In applications $M_{\mathbf{C}}$ describes a model for systems of n components, where for certain subsystems $J \in \mathbf{C}$ one knows the distributions P_J "exactly", i.e. there are many joint measurements of these components available. The marginal problem only arises if the specification of P_J is not exact. Of particular relevance for applications is the _modelling problem_. This means that one should not only solve constructively the marginal problem, but moreover construct submodels $\mathfrak{P} = \{P_\vartheta,\ \vartheta \in \Theta\} \subset M_{\mathbf{C}}$ with the parameter ϑ specifying interesting aspects of the model like e.g. values of certain dependence measures. An interesting problem in this connection is to find an optimal fit of a probability measure (resp. a density) by an element of $M_{\mathbf{C}}$ (resp. a corresponding density with fixed marginals). For several distances characterizations of the optimal fit have been derived (cf. [10], [84], [91]), allowing in some cases explicit resp. "approximative" solutions. Measuring the distance by the Kullback-Leibler measure an iterative procedure, the "iterative proportional fitting" (IPF) resp. "scaling projection method" has been investigated in the literature. But so far only in the finite discrete case a valid convergence proof has been found (cf. [10]). Most papers are concerned with the case $\mathbf{C} = \{\{1\},...,\{n\}\}$ of simple marginals. In this case we use the notation

(2) $M_{\mathfrak{C}} = M(P_1,...,P_n)$.

Some relevant papers on construction problems are [62], [43], [55], [52], [87], [100].

A third class of problems is to find upper and lower bounds for $\int \varphi \, dP$, $\varphi: E \to \mathbb{R}$ measurable, only based on the knowledge of the marginal structure. The optimal bounds are called Fréchet-bounds, defined by:

(3) $M_{\mathfrak{C}}(\varphi) = \sup\{\int \varphi \, dP;\ P \in M_{\mathfrak{C}}\}$, $m_{\mathfrak{C}}(\varphi) = \inf\{\int \varphi \, dP;\ P \in M_{\mathfrak{C}}\}$.

Since $m_{\mathfrak{C}}(\varphi) = -M_{\mathfrak{C}}(-\varphi)$ it is enough to consider either $M_{\mathfrak{C}}$ or $m_{\mathfrak{C}}$. The classical Fréchet-bounds concern the case of simple marginals and $E_i = \mathbb{R}^1$, $1 \le i \le n$. Then $P \in M(P_1,...,P_n)$, if and only if the distribution function $F = F_P$ satisfies:

(4) $\underline{F}(x) \le F(x) \le \overline{F}(x)$, $x \in \mathbb{R}$,

where $\underline{F}(x) = (\sum_{i=1}^{} F_i(x_i) - (n-1))_+$, $\overline{F}(x) = \min(F_i(x_i))$. $\underline{F}, \overline{F}$ are the "lower" resp. "upper" Frechet-bounds. $\overline{F}$ is a distribution function (is an element of the Fréchet-class $\mathfrak{F}(P_1,...,P_n)$), $\underline{F} \in \mathfrak{F}(P_1,...,P_n)$ if $n = 2$, but Dall'Aglio (1972) showed that for $n \ge 3$, $\underline{F}$ is a df only in very exceptional cases. Based on (4) many authors established sharp bounds for $n = 2$ and $\varphi(x,y) = \psi(x-y)$, ψ convex (or concave); in particular $\varphi(x,y) = |x-y|^\alpha$, $\alpha \ge 1$, cf. [30] - [33], [97], [11] - [14], [8], [112], [108], [109], [79], [25]. In particular we refer to the interesting survey article of Dall'Aglio (1972).

More general results on Frechet-bounds can be derived from duality theory. Define the dual problems corresponding to (3)

(5)
$$U(\varphi) := \inf \{ \sum_{J \in \mathfrak{C}} \int f_J \, dP_J;\ \sum_{J \in \mathfrak{C}} f_J \circ \pi_J \ge \varphi \}$$
$$I(\varphi) := \sup \{ \sum_{J \in \mathfrak{C}} \int f_J \, dP_J;\ \sum_{J \in \mathfrak{C}} f_J \circ \pi_J \le \varphi \},$$

then, obviously,

(6) $M_{\mathfrak{C}}(\varphi) \le U(\varphi)$, $I(\varphi) \le m_{\mathfrak{C}}(\varphi)$

and the question of equality in (6) and the existence of solutions is interesting. Some general results on this question were derived in [58], [75], [34], [79], [47], [48], [49], [70], yielding explicit results in particular in the case of simple marginals. In the case of multivariate marginals there are only few papers on Fréchet-bounds resp. Fréchet-classes (cf. [14], [111], [96], [98]).

Applications concern almost sure representations of stochastic orders (Strassen's result), construction of maximally dependent random variables, random variables with maximum sums, r.v.'s with minimum variance of the sum (Monte Carlo Simulation), the Monge-Kantorovic mass transportation problem, construction of minimal metrics and optimal couplings and many others. A basic problem for the study of Fréchet bounds is the study of inequalities of the type $\varphi \leq \sum_{J \in \mathcal{E}} f_J \circ \pi_J$ arising in the definition of the bounds in (5).

We finally mention some statistical problems connected with marginal models. A general question is the following: How can one improve statistical procedures knowing the marginal structure in comparison to the status of ignorance. A different question concerns the robustness of statistical procedures against departures from an ideal independent situation by dependence. There are some close connections between the marginal problem and some recent papers on graphical interaction models, which allow a simplified statistical analysis by their inherent conditional independence properties (cf. [16], [56]). A stochastic ordering result in marginal models allows an easy proof of the Huzurbazar conjecture on partial sufficiency (cf. [95]).

2.   Existence and Duality

One method to prove existence and duality results for the marginal problem is to apply some wellknown duality theorems for (topological) vector spaces. This leads to general duality results, where $M_{\mathcal{E}}(\varphi)$, $m_{\mathcal{E}}(\varphi)$ are replaced by

(7) $\tilde{M}_{\mathcal{E}}(\varphi) = \sup \{ \int \varphi dP; \; P \in M_{\mathcal{E}} \}$,

where $\tilde{M}_{\mathcal{E}} := \mathrm{ba} (P_J, \; J \in \mathcal{E})$ is the set of finite additive contents with marginals P_J. In a second step one has to establish conditions on φ, resp. the topology, to ensure that

(8) $\tilde{M}_{\mathcal{E}}(\varphi) = M_{\mathcal{E}}(\varphi)$, $\tilde{m}_{\mathcal{E}}(\varphi) = m_{\mathcal{E}}(\varphi)$

and to ensure the existence of solutions. This approach has been developed in [75], [76], [79]. The first step can also be based on the Hahn–Banach theorem directly. This has been discussed in greater generality by Lembcke (1972) and Luschgy and Thomsen (1983) (the latter paper also including a discussion on extreme points). The following formulation in Section 2.1 arose from a discussion with H. Luschgy.

2.1.   A Generalization of the Marginal Problem

Let on a general measure space $(X, \mathcal{B})$ (which in this section is not necessarily a product space) $\mathcal{B}_i \subset \mathcal{B}$, $i \in I$, be a system of sub-σ-algebras with probability measures $P_i \in M^1(X, \mathcal{B}_i)$, $i \in I$. Define

(9)
$$M = \{P \in M\ (X, \mathcal{B}); \ P|\mathcal{B}_i = P_i, \ i \in I\}$$
$$\tilde{M} = \{P \in ba(X, \mathcal{B}); \ P|\mathcal{B}_i = P_i, \ i \in I\};$$

$\tilde{M}$ is the set of bounded additive contents with marginals P_i. We assume consistency of (P_i), i.e.

(10) $A \in \mathcal{B}_{i_1} \cap \mathcal{B}_{i_2}$ implies that $P_{i_1}(A) = P_{i_2}(A)$.

Furthermore, we define

(11) $F = \{\sum\limits_{i \in I_o} f_i; \ I_o \subset I \text{ finite}, \ f_i \in \mathcal{L}^1(\mathcal{B}_i, P_i)\} = \bigoplus\limits_{i \in I} \mathcal{L}^1(\mathcal{B}_i, P_i)$

the direct sum of the $\mathcal{B}_i$-measurable functions which are integrable w.r.t. P_i. F is a vector subspace of the vectorspace

(12) $\mathcal{L}^m = \{\varphi \in \mathcal{L}(X, \mathcal{B}); \ \exists f \in F \text{ with } \varphi \leq f\},$

the set of measurable functions which are majorized by an element of F. By consistency the linear operator

(13) $T: F \to \mathbb{R}$, $T(\sum\limits_{i \in I_o} f_i) = \sum\limits_{i \in I_o} \int f_i \, dP_i$

is well defined.

Theorem 1. a) (Marginal Problem)
$\tilde{M} \neq \emptyset$ iff $T \geq 0$ (i.e. $f \in F$, $f \geq 0$ implies $Tf \geq 0$).

b) (Duality) For $\varphi \in \mathcal{L}^m$ we have:

(14) $\tilde{M}(\varphi) := \sup\{\int \varphi \, dP; \ P \in \tilde{M}\} = U(\varphi) := \inf\{Th; \ h \in F, \ \varphi \leq h\}.$

c) If $U(\varphi) > -\infty$, then there exists a $P \in \tilde{M}$ with $\tilde{M}(\varphi) = \int \varphi \, dP$.

Proof. a) The direction "$\Rightarrow$" is trivial. For the converse direction observe that U is sublinear on $\mathcal{L}^m$ and $Uf = Tf$ for $f \in F$. If S is a linear functional on $\mathcal{L}^m$, $S \leq U$, then for $f \in F$, $f \geq 0$, holds: $-Sf = S(-f) \leq U(-f) = \inf\{Th; -f \leq h, h \in F\} \leq T0 = 0$ i.e. $S \geq 0$ and, obviously, $S|F = T$.

By Hahn-Banach there exists an extension S of T to $\mathcal{L}^m$, $S \leq U$. Riesz' representation theorem ensures the existence of an element $P \in ba(X, \mathcal{B})$ representing S. Since $S|F = T$, it follows that $P \in \tilde{M}$.

b), c) A corollary to the Hahn-Banach theorem is the existence of an extension S with $S\varphi = U\varphi$ if $U\varphi > -\infty$. The corresponding content then yields b), c) if $U\varphi > -\infty$. If $U\varphi = -\infty$, then also $M(\varphi) = -\infty$; so b) is valid generally.

Remark. Related existence problems are proved similarly. Let e.g. for a finite measure μ $\tilde{M}_\mu$: = $\{P \in ba(X,\mathcal{B}); P|\mathcal{B}_i = P_i, i \in I, P \leq \mu\}$. Replace the operator from (14) by $U_\mu(\varphi) = \inf \{U(\varphi_o) + \int h_+ d\mu; \varphi_o + h \geq \varphi\}$. Then the existence and duality results analogously to (14) are valid (cf. [57]).

Consider next the following assumptions:

A.1 $(X,\mathcal{B}_i,P_i)$, $i \in I$, are compactly approximable, i.e. there exist compact set-systems $\mathcal{C}_i \subset \mathcal{B}_i$ with $P_i(B_i) = \sup \{P_i(E_i); E_i \subset B_i, E_i \in \mathcal{C}_i\}$, $i \in I$.

A.2 $(X,\mathcal{B})$ is a topological space with Borel σ-algebra $\mathcal{B}$ and $\mathcal{R} = \mathcal{R}(\underset{i \in I}{\cup} \mathcal{B}_i)$ contains a countable basis of the topology.

Let $\mathcal{L}^1(X,\mathcal{R}(\underset{i \in I}{\cup} \mathcal{A}_i), P)$ denote the set of P-integrable functions, where P is considered as a content on the algebra $\mathcal{R}(\underset{i \in I}{\cup} \mathcal{B}_i)$ (cf. Dunford, Schwartz (1967), Def. 17, p. 112).

Theorem 2. a) If A.1 holds, then: $M \neq \emptyset$ iff $T \geq 0$. Furthermore, $M(\varphi) = U$ for $\varphi \in \cap \mathcal{L}^1(X,\mathcal{R}(\underset{i \in I}{\cup} \mathcal{A}_i),P)$.

b) If A.1 and A.2 hold, then $M(\varphi) = U(\varphi)$ for $\varphi \in C_b(X)$.

Proof. A.1 implies that any $P \in \tilde{M}$ is compactly approximable on $\mathcal{R}(\underset{i \in I}{\cup} \mathcal{A}_i)$ and, therefore, σ-additive on $\mathcal{R}(\underset{i \in I}{\cup} \mathcal{A}_i)$, implying the existence of a σ-additive extension. The proof of the duality theorem is similar to [79], [84], Theorem 3. □

Remark. The duality part of Theorem 2 in the case of multivariate marginals was stated in [84], Theorem 3, for upper and lower semicontinuous functions. The indicated proof is only valid for bounded continuous function. It can presumably be extended to upper semi-continuous functions (one has to prove, that U is σ-continuous for increasing sequences), but the result is not true for lower semicontinuous functions, as was indicated by a counterexample of H. Kellerer. □

2.2. The Case of Simple Marginals. $M_\mathcal{C} = M(P_1,...,P_n)$

In the case of simple marginals the duality and existence results of 2.1 have been generalized by Kellerer (1984) to more general functions and spaces. The proofs are based on the study of the continuity properties of $M_\mathcal{C}$, U resp. $m_\mathcal{C}$, I. These continuity properties combined with Choquet's capacity theorem yield in particular the following duality theorem. $(E_i, \mathcal{A}_i)$ are assumed to be Hausdorff topological spaces with Borel σ-algebras $\mathcal{A}_i$. This assumption is made throughout the rest of this paper. Also we assume generally that P_J are Radon measures, $J \in \mathcal{C}$.

Theorem 3. (Kellerer (1984), Theorem 2.21)

The duality theorem $M_{\mathbf{e}}(\varphi) = U(\varphi)$ is true for

a) $\varphi \in \mathbf{\$}$, the class of upper-semicontinuous functios with values in $\bar{\mathbb{R}}$.
        In this case there exists a maximal measure,

b) $\varphi \in \mathbf{\$}_m(E, \otimes \mathbf{\mathfrak{A}}_i)$, the class of measurable functions w.r.t. $\otimes \mathbf{\mathfrak{A}}_i$,
        which are majorized from below by an element of F(F as in Section
        2.1). There exists a solution $f^* \in F$ of the dual problem if $U(\varphi) < \infty$.
        The product σ-algebra can be replaced by the Baire σ-algebra.

c) $\varphi \in S_m(E)$, the class of lower majorized Suslinfunctions w.r.t. $\mathbf{\$}$.   □

Kellerer (1987), Proposition 5.6 also proved a related result for multivariate marginals in the decomposable case.

An interesting consequence of the duality theorem is the following theorem of Strassen (1965) (who gave a proof in the case of polish spaces) saying that for n = 2 one can restrict in the definition of U to two valued functions.

Theorem 4. (Strassen (1963), Kellerer (1984))

Let $n = 2$ and $B \in \mathbf{\mathfrak{A}}_1 \otimes \mathbf{\mathfrak{A}}_2$, then

(15) $M_{\mathbf{e}}(B) = \inf \{P_1(B_1) + P_2(B_2); \ B \subset \bigcup_{i=1}^{2} \pi_i^{-1}(B_i)\}$

               $m_{\mathbf{e}}(B) = \sup \{P_1(B_1) + P_2(B_2) - 1; \ B \supset B_1 \times B_2\}$.

If B is closed, then B_i can be restricted to the class of closed sets.   □

We next discuss some more concrete applications of the duality theorem.

2.2.1. Stochastic Orders

Let $n = 2$, $E_1 = E_2 = Y$, $(Y, \leq)$ an ordered topological space with

(16) $R(Y) = \{(x,y) \in Y \times Y; \ x \leq y\}$

closed, then one obtains from (15) the a.s. representation for the stochastic order $\leq_{s,t}$ w.r.t. monotone increasing functions.

Theorem 5. a) (cf. [47], Prop. 3.11; [85], Lemma 1)

(17) $M_{\mathbf{e}}(R(Y)) = 1 - \sup \{P_1(A) - P_2(A); \ A \text{ closed, isotone}\}$.

b) (Strassen representation theorem, cf. [101])

$$P_1 \leq_{st} P_2 \iff \exists P \in M(P_1, P_2) \text{ with } P(R(Y)) = 1. \qquad \square$$

Strassen's a.s. representation theorem has ben very influential for the theory and applictions of stochastic ordering. It has been extended to ordering results for stochastic processes (cf. [44], [103], [78], it has been extended to "stochastic" ordering not induced by a partial order on Y as e.g. the ordering w.r.t. convex functions (cf. [104] , [77]) and found many applications in the ordering of queues, Markov chains, risk theory and in statistics (cf. [103], [62], [71], [78], [88]).

2.2.2. Sharpness of the Classical Fréchet-Bounds

For product sets $A = A_1 \times \ldots \times A_n \in \mathfrak{A}_1 \otimes \ldots \otimes \mathfrak{A}_n$ there is the obvious generalization of the Fréchet-bounds in (4).

Theorem 6. (cf. [79], Theorem 6)

(18)
$$M_{\mathbf{e}}(A_1 \times \ldots \times A_n) = \min \{P_i(A_i); \ 1 \leq i \leq n\}$$
$$m_{\mathbf{e}}(A_1 \times \ldots \times A_n) = \left(\sum_{i=1}^{n} P_i(A_i) - (n-1) \right)_+ .$$

Proof. In the case $n = 2$ we obtain from (15)

$$M_{\mathbf{e}}(A_1 \times A_2) = \inf \{P_1(B_1) + P_2(B_2); \ A_1 \times A_2 \subset B_1 \times E_2 \cup E_1 \times B_2\}$$
$$= \min (P_1(A_1), P_2(A_2)).$$

The case $n \geq 2$ is proved by induction. Let $Q_0 \in M(P_1, \ldots, P_n)$ satisfy $Q_0(A_1 \times \ldots \times A_n) = \min (P_i(A_i))$, then $\sup \{P(A_1 \times \ldots \times A_{n+1}); \ P \in M(P_1, \ldots, P_{n+1})\} =$

$$\sup_{Q \in M(P_1, \ldots, P_n)} \ \sup_{P \in M(Q, P_{n+1})} P(A_1 \times \ldots \times A_{n+1}) \geq \sup_{P \in M(Q_0, P_{n+1})} P(A_1 \times \ldots \times A_{n+1})$$

$$= \min (Q_0(A_1 \times \ldots \times A_n), P_{n+1}(A_{n+1})) \text{ (from the case } n = 2)$$

$$= \min (P_i(A_i)) \text{ by induction.}$$

Since the oppositive inequality is trivial, the first part of (18) is proved. The proof of the second part is analogously. $\square$

In [79] the proof was given by direct calculation of the upper resp. lower bounds for the duality theorem. Theorem 6 implies that the Fréchet bounds are identical to the Bonferoni bounds in the following sense. Define $B_i := E_1 \times \ldots \times A_i \times \ldots \times E_n$, $p_i = P_i(A_i)$, then $A_1 \times \ldots \times A_n = \bigcap_{i=1}^{n} B_i$. The Bonferoni bounds are defined by

(19) $\tilde{U}((p_i)) = \sup \{P(\bigcap_{i=1}^{n} B_i); \ B_i \in \mathfrak{A}, \ P(B_i) = p_i, \ 1 \leq i \leq n\}$

$\tilde{L}((p_i)) = \inf \{P(\bigcap_{i=1}^{n} B_i); \ B_i \in \mathfrak{A}, \ P(B_i) = p_i, \ 1 \leq i \leq n\}.$

So the knowledge of the whole marginal distribution does not help to obtain better bounds for product sets in comparision to knowing only the probabilities $P_i(A_i)$. For the ordering by survival functions

(20) $P \leq_s Q$ if $P([x,\infty)) \leq Q([x,\infty))$

for all $x \in \mathbb{R}^n$ it has been proved that

(21) $P \leq_s Q$ iff $\int \varphi dP \leq \int \varphi dQ$

for Δ-monotone (in pairs) resp. quasimonotone resp. L-superadditive functions (cf. Cambanis, Simons and Stout (1976) and Whitt (1976) for n = 2, Rüschendorf (1979, 1981, 1983), Tchen (1980), Marshall, Olkin (1979), Mosler (1982) for $n \geq 2$), (21) combined with (18), (4) imply sharp results for $M_{\varphi}(\varphi)$. These are related to rearrangement inequalities (cf. [112], [81]). The case n = 2, $\varphi(x,y) = -\psi(x-y)$, ψ convex, is due to Bertino (1966). Some partial results are in [79] for the lower bound $m_{\varphi}(\varphi)$. An open problem is e.g. to determine $m_{\varphi}(\varphi)$ for Y = [0,1], P_i = R(0,1), the uniform distribution $1 \leq i \leq 3$ and $\varphi(x) = \prod_{i=1}^{3} x_i$. □

2.2.3. Representation of Minimal Metrics

Some of the wellknown probability metrics have a representation as a "minimal metric".

2.2.3.1. Levy-Prohorov-Metric

On a metric space (Y,d) with Borel σ-algebra $\mathfrak{B}$ define for $A \in \mathfrak{B}$, $\varepsilon > 0$,

(22) $A^{\varepsilon} := \{y \in Y; \ d(x,y) < \varepsilon, \ \text{for some} \ x \in A\}, \ A^{\circ} := \overline{A}.$

From (15) we obtain

Theorem 7. (cf. Dudley (1976), Theorem 18.2)

Let $P_1, P_2 \in M^1(Y,\mathfrak{B})$, $\varepsilon \geq 0$;

a) $\underline{\delta \geq 0}$. There exists $P \in M(P_1,P_2)$ with $P\{(x,y) \in Y \times Y; \ d(x,y) > \varepsilon\} < \delta$

$\leftrightarrow \forall A \in \mathfrak{F} = \mathfrak{F}(Y): \ P_1(A) \leq P_2(A^{\varepsilon}) + \delta;$

b) $\underline{\delta \geq 0}$. There exists $P \in M(P_1,P_2)$ with $P\{d(x,y) > \varepsilon\} \leq \delta$

$\leftrightarrow P_1(A) \leq P_2(A^{\varepsilon}) + \delta, \ \forall A \in \mathfrak{F}(Y).$ □

Theorem 7 implies the Strassen-representation of the Levy-Prohorov metric

(23) $\pi(P_1, P_2) = \inf \{\varepsilon > 0: P_1(A) \leq P_2(A^\varepsilon) + \varepsilon, \ \forall A \in \mathcal{B}(Y)\}$.

Define for $P \in M^1(Y \times Y, \mathcal{B} \otimes \mathcal{B})$ the Ky-Fan (probability-) metric

(24) $K(P) = \inf \{\varepsilon > 0: P(d(x,y) > \varepsilon) < \varepsilon\}$

and consider the corresponding minimal metric

(25) $\hat{K}(P_1, P_2) = \inf \{K(P); \ P \in M(P_1, P_2)\}$.

Theorem 8. (Strassen (1964), Dudley (1968))

(26) $\hat{K} = \pi$.   □

π metrizes the topology of weak convergence on the set of tight Borel measures (this is immediate from Dudley (1976), Theorem 8.3, who co siders the case of separable metric spaces), i.e.

(27) $P_n \overset{\mathcal{D}}{\longrightarrow} P$ if and only if $\pi(P_n, P) \to 0$.

A basic coupling result is the almost sure representation theorem. The proof of part b) makes essential use of Theorem 8.

Theorem 9. (Almost sure representation theorem)

a)   (Skorohod, Strassen, Dudley, Wichura, cf. [23])

Let $P_n, P \in M^1(Y,d)$ be tight. Then $\pi(P_n, P) \to 0$ if and only if there exists a probability space $(\Omega, \mathcal{A}, R)$ and Y-valued random variables X_n, X on $(\Omega, \mathcal{A})$, such that $R^{X_n} = P_n$, $R^X = P$ and $d(X_n, X) \to 0$ a.s.

b)   Rachev, Rüschendorf and Schief (1988), Dudley (1989)

If $P_n, Q_n \in M^1(Y,d)$ are tight and $\pi(P_n, Q_n) \to 0$, then $d(X_n, Y_n) \to 0$ a.s. for some versions X_n, Y_n on a probability space $(\Omega, \mathcal{A}, R)$ with $R^{X_n} = P_n, R^{Y_n} = Q_n$.   □

Remark. If $P_n \in M(Y)$, $Y = Y_1 \times Y_2$ a product space, $P_n \in M(Q_n, R)$ and $P_n \overset{\mathcal{D}}{\longrightarrow} P$, then the following sharpening of Theorem 8 is not true: "There exist versions of P_n of the form (X_n, Z), such that (X_n) is a.s. convergent (cf. [70])."   □

2.2.3.2. $\mathcal{S}^p$-Metrics

Define the probability metric

(28) $\qquad \mathbf{\mathit{g}}^{\infty}(P) = \text{ess}\sup_{P} d(x,y) = \inf\{\epsilon > 0; \; P(d(x,y) > \epsilon) = 0\}.$

Theorem 7.b) implies the following representation of the corresponding minimal metric.

Theorem 10. (cf. Dudley (1976), Theorem 18.2)

(29) $\qquad \widehat{\mathbf{\mathit{g}}^{\infty}}(P_1,P_2) := \inf\{\mathbf{\mathit{g}}^{\infty}(P); \; P \in M(P_1,P_2)\}$

$\qquad\qquad = \inf\{\epsilon > 0; \; P_1(A) \le P_2(A^\epsilon), \; \forall A \in \mathbf{\mathit{g}}(Y)\}.$　□

For the $\mathbf{\mathit{g}}^p$-distance, $1 \le p < \infty$

(30) $\qquad \mathbf{\mathit{g}}^p(P) = \int d^p(x,y)\,dP(x,y)$

(the corresponding probability metric is $d_p(P) = (\mathbf{\mathit{g}}^p(P))^{1/p}$) the duality theorem of Section 2.2 implies for $P_1, P_2 \in M^1(Y)$ with $\int d^p(x,a)dP_1(x) < \infty$

(31) $\qquad \widehat{\mathbf{\mathit{g}}^p}(P_1,P_2) = \sup\{\int f\,dP_1 + \int g\,dP_2; \; f \in \mathbf{\mathit{g}}(P_1), \; g \in \mathbf{\mathit{g}}(P_2), \; f(x)$

$\qquad\qquad + g(y) \le d^p(x,y)\}$

(cf. also [68]). For $p = 1$ there is the following strengthening of (31).

Theorem 11. (Kantorovic-Rubinstein-Theorem, cf. [46], [58], [106], [107], [28], [47], [70]).

If $\int d(x,a)(P_1 + P_2)(dx) < \infty$, then

(32) $\qquad \widehat{\mathbf{\mathit{g}}^1}(P_1,P_2) = \sup\{\int f\,d(P_1 - P_2); \; f \in \text{Lip}(Y)\}, \quad$ where

$\qquad \text{Lip}(Y) = \{f: Y \to \mathbb{R}^1; \; |f(x) - f(y)| \le d(x,y)\}.$

Proof. From (31)

$\qquad \widehat{\mathbf{\mathit{g}}^1}(P_1,P_2) = \sup\{\int f_1\,dP_1 - \int f_2\,dP_2; \; f_1(x) - f_2(y) \le d(x,y), \; f_i \in \mathbf{\mathit{g}}^1(P_i)\}.$

Let $P_i = R + R_i$, $i = 1,2$, be a decomposition of P_i, $i = 1,2$, where the measures R_i are orthogonal with supports A_1, A_2. If $R = 0$, then define

$$f(x) := \begin{cases} \sup\{f_1(x_1) - d(x_1,x); \; x_1 \in A_1\} & \text{if } x \in A_2 \\ \inf\{f_2(x_2) + d(x_2,x); \; x_2 \in A_2\} & \text{if } x \in A_1 \end{cases}$$

Then $f \in \text{Lip}(Y)$ and f is better then f_1, f_2, i.e. $f(x) \ge f_1(x)$, $x \in A_1$, $f(x) \le f_2(x)$, $x \in A_2$, and, therefore, $\int f_1\,dP_1 - \int f_2\,dP_2 = \int_{A_1} f_1\,dR_1 - \int_{A_2} f_2\,dR_2 \le \int f\,d(R_1 - R_2) = \int f\,d(P_1 - P_2)$ and from (31) $\mathbf{\mathit{g}}^1(P_1,P_2) = I(P_1,P_2) := \sup\{\int f\,d(P_1 - P_2); f \in \text{Lip}(Y)\}$. If $R \ne 0$, $R_i \ne 0$, then the result follows from the following relations:

$\widehat{\mathbf{\mathit{g}}^1}(P_1,P_2) \ge I(P_1,P_2) = I(R_1,R_2) = \widehat{\mathbf{\mathit{g}}^1}(R_1,R_2) := \inf\{\int d(x,y)d\tilde{R}(x,y), \tilde{R} \in M(R_1,R_2)\}$

$\ge \inf\{\int d(x,y)dP(x,y); \; P \in M(P_1,P_2)\} = \widehat{\mathbf{\mathit{g}}^1}(P_1,P_2)$. For the last inequality observe

that for $\tilde{R} \in M(R_1, R_2)$ and $Q \in M(R,R)$ concentrated on the diagonal,
$P := Q + \tilde{R} \in M(P_1, P_2)$ and $\int dx,y)dP = \int d(x,y)d\tilde{R}$. If $R_I = 0$, then $P_1 = P_2$ and the equality is trivial.   □

The idea of this proof is due to Szulga (1978). The integrability assumption $\int d(x,a)(P_1 + P_2)(dx) < \infty$ was removed by Kellerer (1984). For $Y = \mathbb{R}^1$ the minimal $\mathfrak{L}^p$ metrics are explicitly known (cf. Gini [36], Salvemini [97], Dall'Aglio [11] - [14], Fréchet [30] - [33], Hoeffding [40], Vallender [109]).

(33) $\widehat{\mathfrak{L}^1}(P_1, P_2) = \int |F_1(x) - F_2(x)|dx$

$\widehat{\mathfrak{L}^p}(P_1, P_2) = \int_0^1 |F_1^{-1}(t) - F_2^{-1}(t)|^p dt, \ p \geq 1$,

where F_I are the df's of P_I.

For $Y = \mathbb{R}^k$ there are few explicit solutions. Let $|x|$ denote the euclidean norm in $\mathbb{R}^k$.

Theorem 12. a) (Knott and Smith (1984), Rüschendorf and Rachev (1990))
If $\int |x|^2 dP_i(x) < \infty$, $i = 1,2$, then random variables X, Y with distributions P_1, P_2 satisfy: $E|X - Y|^2 = \widehat{\mathfrak{L}^2}(P_1, P_2) \Leftrightarrow \exists \ f: \mathbb{R}^k \to \mathbb{R}^1$ closed, convex such that a.s. $Y \in \partial f(X)$, the subgradient of f in X.

b) (Dawson and Landau (1982), Olkin and Pukelsheim (1982), Givens and Shortt (1984))

For nonsingular covariance matrices Σ_1, Σ_2 and $a_1, a_2 \in \mathbb{R}^k$:

(34) $\widehat{\mathfrak{L}^2}(N(a_1, \Sigma_1), N(a_2, \Sigma_2)) = |a_1 - a_2|^2 + \text{tr} \, \Sigma_1 + \text{tr} \, \Sigma_2 - 2 \, \text{tr}(\Sigma_1^{1/2} \Sigma_2 \Sigma_1^{1/2})^{1/}$

Remark.

a)   A differentiable continuous function f is convex, if and only if

(35) $\Phi(x) := \nabla f(x)$ is monotone, i.e. $\langle x - y, \Phi(x) - \Phi(y) \rangle \geq 0, \ \forall x,y$

(cf. [74], p. 99). Therefore, if X has distribution P_1, if Φ is the gradient of a differentiable function f, $\Phi(X) = \nabla f(X)$, has distribution P_2, then $(X, \Phi(X))$ is an optimal coupling w.r.t. $\mathfrak{L}^2$-distance if and only if Φ is monotone. Let $\omega = \sum_{i=1}^k \Phi_i \, dx_i$, then $d\omega = \sum_{i<j} (\frac{\partial \Phi_j}{\partial x_i} - \frac{\partial \Phi_i}{\partial x_j}) dx_i \wedge x_j$ and by Poincaré's lemma we obtain: If Φ is continuously differentiable, then $(X, \Phi(X))$ is an optimal coupling w.r.t. $\mathfrak{L}^2$-distance, if and only if

(36) 1. $\quad \dfrac{\partial \Phi_j}{\partial x_i} = \dfrac{\partial \Phi_i}{\partial x_j}$, $\forall i \neq j$ and

2. Φ is monotone.

For linear functions $\Phi(x) = Ax$, this is equivalent to the assumption that A is positive semidefinite and symmetric. In the normal case (34) with $a_1 = a_2 = 0$, we obtain with $\Phi(x) = \Sigma_1^{-1/2} \Sigma_2^{1/2} x$: $(X, \Phi(X))$ is optimal if $\Sigma_1 \Sigma_2 = \Sigma_2 \Sigma_1$. In the general case we use $\Phi(x) = \Sigma_1^{1/2} (\Sigma_1^{1/2} \Sigma_2 \Sigma_1^{1/2})^{-1/2} \Sigma_1^{1/2} x$ to obtain (34).

b) The proof of Theorem 12. a) can be based on the duality theorem and results from convex analysis (for some extensions cf. Section 3). The minimal ℓ^p-metrics are special instances of the Monge-Kantorovic mass transference problem. A review of this type of problems and several further representations of minimal metrics are given in Rachev (1985). Some results for the ℓ^2-metrics and their application to approximation problems are discussed in [86].

2.2.4. Probability on Diagonals

For $n \geq 2$, $E_1 = \ldots = E_n = Y$ and for $A \in \mathfrak{A}_1$ let

(37) $\quad \Delta_n(A) := \{(x,\ldots,x); \ x \in A\} \in \overset{n}{\underset{i=1}{\otimes}} \mathfrak{A}_1$

and let

(38) $\quad P_1 \wedge \ldots \wedge P_n(A) := \inf \{ \sum\limits_{i=1}^{n} P_i(A_i); \ A_i \in \mathfrak{A}_1, \ \sum\limits_{i=1}^{n} A_i = A\}.$

<u>Theorem</u> 13. $\exists P^* \in M(P_1,\ldots,P_n)$ such that for all $A \in \mathfrak{A}_1$:

(39) $\quad P^*(\Delta_n(A)) = M_{\oplus}(\Delta_n(A)) = P_1 \wedge \ldots \wedge P_n(A).$

<u>Proof.</u> For $n = 2$ the equality $M_{\oplus}(\Delta_2(A)) = P_1 \wedge P_2(A)$ follows from (15). Since $P_1 \wedge P_2$ is a measure on Y, $M_{\oplus}$ is additive on $\mathfrak{A}_1$. This implies by an inductive argument the existence of $P^* \in M(P_1,P_2)$ with $P^*(\Delta_2(A)) = P_1 \wedge P_2(A)$, $\forall A \in \mathfrak{A}_1$ (If $P^*(A_1 + A_2) = M_{\oplus}(A_1 + A_2) = M_{\oplus}(A_1) + M_{\oplus}(A_2)$, then $P^*(A_i) = M_{\oplus}(A_i)$, $i = 1,2)$.

If $n \geq 2$, then take $Q \in M(P_1,\ldots,P_{n-1})$ with $Q(\Delta_{n-1}(A)) = P_1 \wedge \ldots \wedge P_{n-1}(A)$, $A \in \mathfrak{A}_1$, (induction hypothesis). Then $M_{\oplus}(\Delta_n(A)) = \underset{Q' \in M(P_1,\ldots,P_{n-1})}{\sup} \underset{P \in M(Q',P_n)}{\sup} P(\Delta_n(A))$

$\geq \underset{P \in M(Q,P_n)}{\sup} P(\Delta_n(A)) = \inf \{Q(B_1) + P_n(B_2); \ \Delta_n(A) \subset B_1 \times Y \cup Y \times B_2\}$

$= \inf \{Q(\Delta_n(A_1)) + P_n(A \setminus A_1); \ A_1 \subset A\} = \inf \{P_1 \wedge \ldots \wedge P_{n-1}(A_1) + P_n(A \setminus A_1); \ A_1 \subset A\}$

$= P_1 \wedge \ldots \wedge P_n(A)$. The oppositive inequality is trivial. $\quad\square$

In the case $n = 2$, $A = Y$, (39) implies with $\Delta_2 = \Delta_2(Y)$ and

(40) $d_v(P_1, P_2) = \sup \{P_1(B) - P_2(B); B \in \mathfrak{A}_1\}$,

the wellknown representation of the sup-metric d_v:

(41) $d_v(P_1, P_2) = 1 - M_{\mathfrak{C}}(\Delta_2) = m_{\mathfrak{C}}(\Delta_2)$

due to Dobrushin (1969).

2.2.5. Random Variables With Maximum Sums

Problem: For $P_i \in M(\mathbb{R})$ with df's F_i, $1 \le i \le n$, determine the maximum resp. minimum probability of

(42) $A_n(t) := \{x \in \mathbb{R}^n; \sum_{i=1}^{n} x_i \le t\}$.

This problem was solved independently by Makarov (1981) and Rüschendorf (1982) for $n = 2$. (For a different proof cf. also Frank, Nelsen and Schweizer (1987).) Makarov introduces this problem as "Kolmogorov's problem".

Theorem 14. (Makarov (1981), Rüschendorf (1982))

For $n = 2$ and $t \in \mathbb{R}$ we have

(43) $M_{\mathfrak{C}}(A_2(t)) = F_1 \wedge F_2(t) = \inf_x (F_1(x-) + F_2(t-x))$ the infimal convolution.

(44) $m_{\mathfrak{C}}(A_2(t)) = F_1 \vee F_2(t-) - 1$, where $F_1 \vee F_2(t)) = \sup_x (F_1(x-) + F_2(t-x))$.

                                                                                    □

For $n \ge 2$ there are some particular results in [77], obtained by explicit solution of the dual problem. If e.g. $P_1 = \ldots = P_n = R(0,1)$, then

(45) $M_{\mathfrak{C}}(A_n(t)) = \frac{2}{n} t$, $0 \le t \le \frac{n}{2}$,

(46) $m_{\mathfrak{C}}(A_n(t)) = \min \{(\frac{2}{n} t - 1)_t, 1\}$, $t \ge 0$.

If $P_i = \mathfrak{B}(1, \vartheta)$, $1 \le i \le n$, then

(47) $M_{\mathfrak{C}}(A_n(k)) = \frac{n}{n-k} (1 - \vartheta)$, $k \le n\vartheta$,

the solution $P^* \in M_{\mathfrak{C}}$ being a mixture of the uniform distribution on $\{x: \sum_{i=1}^{n} x_i = k\}$ and a one point measure in $(1, \ldots, 1)$.

Similar formulas are possible for other geometric objects like circles or triangles (for $n = 2$).

2.2.6. Monte-Carlo-Simulation

Problem: For $P_i \in M^1(\mathbb{R}^1)$ construct rv's $X_i^* \sim P_i$ with

(48) $\mathrm{Var} (\sum_{i=1}^{n} X_i^*) \le \mathrm{Var} (\sum_i X_i)$ for any $X_i \sim P_i$.

For n = 2 a solution is the wellknown method of "antithetic variates" (cf. Hammersley Handscomb (1964)). For $n \geq 2$ there are some particular results.

1.   If $P_i = R(0,1)$, then it is possible to construct $X_i^* \sim P_i$, $1 \leq i \leq n$, with $\sum_{i=1}^{n} X_i^* = \frac{n}{2}$ (cf. Gaffke and Rüschendorf (1981)). So (X_i) solve (48) trivially.

2.   If P_i are uniform on $\{1,...,n\}$, then one can construct X_i^*, $1 \leq i \leq n$, with $\sum_{i=1}^{n} X_i^* \in \{a, a+1\}$, which solve (48) (cf. [82]).

3.   If $P_i = \mathcal{B}(1,\vartheta)$, then one can again construct $X_i^* \sim P_i$ with $\sum_{i=1}^{n} X_i^* \in \{k, k+1\}$, $\frac{k}{n} \leq \vartheta \leq \frac{k+1}{n}$, which solve (48). The minimal value of the variance equals the cyclic function

(49) $v_k(\vartheta) = a(k,\vartheta)(1 - a(k,\vartheta))$, $a(k,\vartheta) := k\vartheta \pmod{1}$

(cf. Snijders (1984)).

In these examples it is possible to concentrate the distribution of ΣX_i^* "close" to $n E X_1^*$. For a symmetric distribution (like $N(a, \sigma^2)$) and $n = 2m$ one can choose rv's X_i with $\sum_{i=1}^{n} X_i = n E X_1$.

For $P_i \in M^1(\mathbb{R}^k, \mathcal{B}^k)$, $1 \leq i \leq n$, we can similarly consider $\frac{1}{n} \sum_{i=1}^{n} X_i$ as simulation for $a = \frac{1}{n} \sum_{i=1}^{n} E X_i$ (typically: $P_1 = ... = P_n$) with error $E|\frac{1}{n} \Sigma X_i - a|^2$. The corresponding problem is to determine the minimum of $\sum_{i<j} E \langle X_i, X_j \rangle$. For n = 2 we obtain a characterization of a solution from Theorem 12 (cf. also Section 3.1): $E \langle X_1^*, X_2^* \rangle = \min_{X_1 \sim P_1, X_2 \sim P_2} E \langle X_1, X_2 \rangle$

$\leftrightarrow \exists f : \mathbb{R} \rightarrow \mathbb{R}$ closed, convex, such that $X_2^* \in \partial f(-X_1^*)$.

2.2.7. Maximally Dependent Random Variables

Lai and Robbins (1978) constructed for given $P_i \in M^1(\mathbb{R}^1)$, $i \in \mathbb{N}$, random variables $X_i^* \sim P_i$ such that

(50) $P^{\max_{1 \leq i \leq n} X_i} \leq_{st} P^{\max_{1 \leq i \leq n} X_i}$, $\forall n \in \mathbb{N}$,

$X^* = (X_i)$ is called maximally dependent sequence. In the case $P_i = R(0,1)$ there is a nice geometric construction (cf. also [76]). In terms of limit theorems Lai and Robbins established that $\max_{1 \leq i \leq n} X_i^*$ is not much larger than $\max \tilde{X}_i$, where $(\tilde{X}_i)$ is an independent sequence (in the case $P_i = P_1$, $\forall i$). For a construction based on duality theory cf. [34] , [50]. From (18) one obtains $P(\max\limits_{1 \le i \le n} X_i^* \le t) = (\sum\limits_{i=1}^{n} F_{P_i}(t) - (n-1))_+$. Solutions then can be defined iteratively.

In the case n = 2, P_i = R(0,1), i = 1,2, $M(P_1,P_2)$ is called the class of doubly stochastic measures. Let U be a R(0,1)-distributed random variable and for a λ^1-preserving transformation g: [0,1] → [0,1] define P_g to be the distribution of (U,g(U)), $_gP$ to be the distribution of (g(U),U). If g is one to one then $_gP,P_g$ are called permutation measures since $P_g(A \times B) = \lambda^1(A \cap g^{-1}(B))$, $A,B \in \mathcal{B}^1[0,1]$ and $_gP = P_{g^{-1}}$.

The only monotonic transformations of [0,1] which are λ^1 preserving are $g_1(u) = u[\lambda^1]$, $g_2(u) = 1 - u[\lambda^1]$, the corresponding permutation measures are the Fréchet-distributions. The property of two random variables X,Y that Y = g(X), g λ^1-preserving was introduced by Lancaster (1963) under the notation: Y is <u>completely dependent</u> on X. By Theorem 1 of Brown (1966), $M(P_1,P_2)$ is the closure of the set of all permutation measures w. r.t. weak operator topology on L^1, i.e. w.r.t. convergence of integrals of functions $f(x) \cdot g(y) \in L^1(\lambda^2)$. This theorem implies in particular that each doubly stochastic measure (also the product measure) can be approximated w.r.t. convergence in distribution by a sequence of permutation measures and it is easy to give an explicit constructon of an approximation sequence (cf. also Kimeldorf and Sampson (1978)). So in a certain sense complete dependence is close to independence. This is related to the generation of chaotic (stochastic) behaviour of dynamical systems by deterministic models.

2.3. The Case of Multivariate Marginals

In the case of multivariate marginals there are few explicit results. In the decomposable case there is an interesting reduction principle which is proved in [96] for Borel spaces (the proof being valid for universally measurable separable metric spaces). Let $h_i : E_i \to W_i$ be measurable, E_i, W_i Borel spaces, 1 ≤ i ≤ n, let $h_J := (h_j)_{j \in J} : \prod\limits_{j \in J} E_j \to \prod\limits_{j \in J} W_j$, $J \subset \{1,...,n\}$, h = $(h_1,...,h)$

Theorem 15. (cf. [93])

If $\mathcal{E}$ is decomposable, then

(51) $M_{\mathcal{E}}^h = \{P^h; \ P \in M_{\mathcal{E}}\} = M(P_J^{h_J}; \ J \in \mathcal{E})$.   □

For the special case $M(P_1,...,P_n)^h = M(P_1^{h_1},...,P_n^{h_n})$ cf. also Rachev and Rüschendorf (1986), Scarsini (1989). If $h_i: ([0,1],\lambda) \to (E_i,\mathcal{U}_i,P_i)$ with $(\lambda^1)^{h_i} = P$

then any $P \in M(P_1,...,P_n)$ has a representation $P = Q^h$, $Q \in M(R(0,1),...,R(0,1))$. If $E_i = \mathbb{R}$, $h_i(x_i) = F_i(x_i)$, $x_i \in (0,1)$, where F_i are the df's of P_i, then $\lambda^{h_i} = P_i$ and $P = Q^h$. Therefore,

(52) $\qquad F_P(x) = Q(h \le x) = Q(F_i^{-1} \le x_i, 1 \le i \le n) = F_Q(F_1(x_1),...,F_n(x_n))$,

F_Q is the so called "copula".

(51) implies in particular for the case of simple marginals and $h_i : E_i \to \mathbb{R}^1$, $h_i \in \mathcal{L}^1(P_i)$:

(53) $\qquad M_{\mathcal{C}}(\prod_{i=1}^n h_i) = \int_0^1 \prod_{i=1}^n F_{h_i}^{-1}(u)\,du$,

where F_{h_i} is the df. of $P_i^{h_i}$.

For some decomposable cases in [14], [96] sharp bounds have been proved as e.g. for star-configurations $\mathcal{C}_1 = \{\{1,j\}, 2 \le j \le n\}$ or simple series-configurations $\mathcal{C} = \{\{1,2\}, \{2,3\}\}$. In [96] is a discussion of two principles of deriving bounds, the method of Bonferoni-type bounds and the method of conditioning.

In the nonregular case the set $M_{\mathcal{C}}$ can be empty, can contain one element (uniqueness) or can be a large convex set. This is in contrast with the decomposable case. A further difference is the fact that the continuity properties of $M_{\mathcal{C}}$, U in the nonregular case seem to be strictly weaker than in the regular case. But these properties need a more detailed investigation.

Example. Let $n = 3$, $\mathcal{C} = \{\{1,2\}, \{2,3\}, \{1,3\}\}$, the simplest nonregular case, and let $E_i = [0,1]$. If $P_{ij} = P^{(U,1-U)}$ for all i,j, where $P^U = R(0,1)$ is uniform on $(0,1)$, then $M_{\mathcal{C}} = \emptyset$.

If $P \in M([0,1]^3)$ with marginals (P_{ij}), $i,j \le 3$, and $P\{x : \Sigma x_i = c\} = 1$, then $M_{\mathcal{C}} = \{P\}$. For the proof note that for any $Q \in M_{\mathcal{C}}$ we have $\int(\sum_{j=1}^3 x_j - c)\,dQ(x) = \int(\Sigma x_j - c)\,dP(x) = 0$ i.e. $Q\{x : \Sigma x_j = c\} = 1$. This implies that the conditional distributions $P^{\pi_3|\pi_1 = x_1, \pi_2 = x_2} = Q^{\pi_3|\pi_1 = x_1, \pi_2 = x_2}$ and, therefore, $P = Q$.

If $P_{ij} = R(0,1) \otimes R(0,1)$ for all i,j, then $\lambda^3_{[0,1]} = R(0,1) \otimes R(0,1) \otimes R(0,1) \in M_{\mathcal{C}}$. Let $v_i : [0,1] \to [-1,1]$ satisfy $\int v_i(x)dx = 0$, $1 \le i \le 3$. The measures $P_v := (1 + \Pi v_i(x))\lambda^3_{[0,1]}$, $v = (v_i)$, all have two dimensional marginals $\lambda^2_{[0,1]}$, i.e. $P_v \in M_{\mathcal{C}}$. One can explicitly construct all elements of $M_{\mathcal{C}}$ which are continuous w.r.t. $\lambda^3_{[0,1]}$ (cf. [87]). $\qquad \square$

For $P_{ij} \in M^1(E_i \times E_j)$ let $P_{i|x_j} = P_{ij}^{\pi_i|\pi_j = x_j}$ be the conditional distribution and define for $A \in \mathfrak{A}_1 \otimes \mathfrak{A}_3$

(54)
$$U_{13|x_2}(A) := \inf \{P_{1|x_2}(A_1) + P_{3|x_2}(A_3); \ A \subset A_1 \times E_3 \cup E_1 \times A_2\}$$

$$L_{13|x_2}(A) := \sup \{P_{1|x_2}(A_1) + P_{3|x_2}(A_3); \ A \supset A_1 \times A_3\}.$$

Theorem 16. (cf. [96])

a)   If $\mathfrak{E} = \{\{1,2\}, \{2,3\}\}$, $B \in \mathfrak{B}_1 \otimes \mathfrak{B}_2 \otimes \mathfrak{B}_3$, then

(55)
$$M_{\mathfrak{E}}(B) = \int U_{13|x_2}(B_{x_2}) dP_2(x_2)$$

$$m_{\mathfrak{E}}(B) = \int L_{13|x_2}(B_{x_2}) dP_2(x_2).$$

b)   If $\mathfrak{E} = \{\{1,2\}, \{1,3\}, \{2,3\}\}$, then

(56)
$$M_{\mathfrak{E}}(\varphi) \leq U(\varphi) := \min \{ \int U_{23|x_1}(\varphi_{x_1}) dP_1(x_1), \ \int U_{13,x_2}(\varphi_{x_2}) dP_2(x_2),$$

$$\int U_{12|x_3}(\varphi_{x_2}) dP_3(x_3)\}$$

for $\varphi \in \mathfrak{B}_m(E)$, where $U_{ij|x_k}(\varphi_{x_k})$ are defined analogously to (54). $\square$

From (56) for $\varphi = 1_A$, $A = A_1 \times A_2 \times A_3$ follows

(57)
$$M_{\mathfrak{E}}(A) \leq \tilde{U}(1_A) \leq \min (P_{ij}(A_i \times A_j)),$$

the right hand side being the Bonferoni bound. The last inequality typically is strict. This is in contrast to the case of simple marginals.

In the case $\mathfrak{E} = J_2^3 := \{\{1,2\}, \{1,3\}, \{2,3\}\}$ let

(58)
$$C(P_{12}, P_{23}) = \{P_{13}; \ M(P_{12}, P_{13}, P_{23}) \neq \emptyset\}$$

be the compatibility set of P_{12}, P_{23}. Dall'Aglio (1959, 1972) proved that in the case $E_i = \mathbb{R}^1$:

(59)
$$\underline{F}_{13}(x_1, x_3) := \int \max (F_{1|x_2}(x_1) + F_{3|x_2}(x_3) - 1, 0) dP_2(x_2)$$

$$\leq F_{13}(x_1, x_3) \leq \overline{F}_{13}(x_1, x_3) = \int \min (F_{1|x_2}(x_3), F_{3|x_2}(x_3)) dF_2(x_2),$$

$\underline{F}_{13}$, $\overline{F}_{13}$ are the minimal and maximal df's of elements of $C(P_{12}, P_{23})$. For the converse Dall'Aglio (1959) gives a counterexample.

The following result gives a characterization of the marginal problem $\mathfrak{E} = I_2^3$.

Theorem 17. (cf. [96])

$P_{13} \in C(P_{12}, P_{23}) \leftrightarrow \forall \varphi = \varphi(x_1, x_3)$ bounded, measurable:

(60) $\tilde{L}_{13}(\varphi) \leq \int \varphi \, dP_{13} \leq \tilde{U}_{13}(\varphi)$, where

$$\tilde{U}_{13}(\varphi) = \int U_{13|x_2}(\varphi) dP_2(x_2), \quad \tilde{L}_{13}(\varphi) = \int L_{13|x_2}(\varphi) dP_2(x_2). \qquad \square$$

It is not sufficient to consider indicator functions only.

3.   Inequalities of the Type: $c(x,y) \leq f(x) + g(y)$

In this section motivated by the duality theorem we investigate gene-ralizations of the Young inequality (cf. Section 3.2 for a statement).

3.1.   c-Convex Functions

For $n = 2$, $\mathfrak{C} = \{\{1\}, \{2\}\}$, $P_1, P_2 \in M(Y)$ and $c = c(x,y) \in \mathfrak{B}_m$ from Theorem 3 follows:

(61) $M_{\mathfrak{C}}(c) = \inf \{\int f \, dP_1 + \int g \, dP_2; \ f \in \mathfrak{B}^1(P_1), \ g \in \mathfrak{B}^1(P_2), \ c(x,y) \leq f(x) + g(y)\}$

and there exist solutions of the dual problem, if $\int c(x,a)(P_1 + P_2)(dx) < \infty$. A "maximal" measure $P \in M_{\mathfrak{C}}$ exists if $c \in \mathfrak{B}(Y \times Y)$ i.e. c is upper semi-continuous.

If (f,g) are admissible (i.e. $f(x) + g(y) \geq c(x,y)$) and $P \in M_{\mathfrak{C}}$, then $(P,(f,g))$ are solutions if and only if

(62) $c(x,y) = f(x) + g(y)$ [P].

Therefore, for the calculation of the Frechet-bounds one needs sharp ine-qualities of the type $c(x,y) \leq f(x) + g(y)$.

For $c(x,y) = \pm \langle x,y \rangle$, $x,y \in \mathbb{R}^k$, a theory of these inequalities has been established in the convex conjugate duality theory (cf. Rockafellar (1970)). This led in Theorem 12 to a characterization of optimal couplings w.r.t. $\mathfrak{B}^2$-distance. For general $c: E_1 \times E_2 \to \mathbb{R}^1$, there are several papers, but the results are less complete. For the literature we refer to [19], [27], [42], [1].

For $f: E_1 \to \mathbb{R}^1$ define the c-conjugate

(63) $f^*: E_2 \to \overline{\mathbb{R}}^1$, $f^*(y) = \sup_{x \in E_1} (c(x,y) - f(x))$

and the doubly c-conjugate

(64) $f^{**}: E_1 \to \overline{\mathbb{R}}^1$, $f^{**}(x) = \sup_{y \in E_2} (c(x,y) - f^*(y))$.

Then, for any admissible pair (f,g) we have:

(65) $f(x) + g(y) \geq f(x) + f^*(y) \geq f^{**}(x) + f^*(y) \geq c(x,y).$

Define the <u>equality domains</u> of (f,f^*) by

(66) $E_c f(x) = \{y;\ f(x) + f^*(y) = c(x,y)\}$

                         $E_c f^*(y) = \{x;\ f(x) + f^*(y) = c(x,y)\}.$

Define the class of <u>c-convex</u> functions

(67) $\Gamma^c(E_1) = \{h: E_1 \to \bar{R};\ h(x) = \sup\limits_{i \in I} [c(x,y_i) + a_i]$ for some $a_i \in \bar{R},$
                         $y_i \in E_2,\ i \in I\}$

                         $\Gamma^c(E_2) = \{h: E_2 \to \bar{R};\ h(y) = \sup\limits_{i \in I} [c(x_i,y) + b_i],\ b_i \in \bar{R},\ x_i \in E_1,$
                         I any index set}.

Elster and Nehse (1974) proved that

a) $f^* \in \Gamma^c(E_2),\ f^{**} \in \Gamma^c(E_1),$

b) f^{**} is the largest c-convex function which is majorized by f,

c) $f = f^{**} \iff f \in \Gamma^c(E_1).$

If $c(x,y) = \langle x,y \rangle,\ x \in Y = E_1$ a locally convex topological vector space, $y \in Y^* = E_2,$ then $\Gamma^c(E_1)$ is identical to the class of convex, closed (= lower semicontinuous) functions on Y. From (64) it is clear that in the duality theorem (61) we can restrict to c-convex functions. It is however known that for certain classes of functions the class of c-convex functios is very large, so that in these cases the reduction is not very interesting (cf. [19], [1]).

Theorem 18. For $c \in \mathbf{g}_m$ with $\int c(x,a)dP_i(x) < \infty,\ i = 1,2,$ we have: $P \in M_{\mathbf{g}}$ is a maximal measure induced by random variables $X \sim P_1,\ Y \sim P_2,$ if and only if

(68) $Y \in E_c f(X)$ a.s. for some c-convex $f \in \mathbf{g}^1(P_1)$ or, equivalently, if and only if $X \in E_c f^*(Y).$

Proof. If $Y \in E_c f(X)$ a.s. for some c-convex $f \in \mathbf{g}^1(P_1),$ then for any random variables $\tilde{X} \sim P_1,\ \tilde{Y} \sim P_2$ we have: $E c(\tilde{X},\tilde{Y}) \leq Ef(\tilde{X}) + Ef^*(\tilde{Y}) = E(f(X) + f^*(Y)) = Ec(X,Y)),$ i.e. (X,Y) is an optimal coupling.

There exists a solution (f,g) of the dual problem, $f \in \mathbf{g}^1(P_1),$ $g \in \mathbf{g}^1(P_2).$ By (65) we can w.l.g. assume that f is c-convex and $g = f^*.$ The converse direction is implied by (62).   □

From (68) it is of interest to characterize the equality sets of c-convex functions. For $\bar{f}: E_1 \to R\ \varepsilon > 0,$ define the <u>ε-c-subdifferential</u>

(69) $\partial_{c,\varepsilon} f(x) = \{y: f(x') - f(x) \geq c(x',y) - c(x,y) - \varepsilon,\ \forall x' \in E_1\},$

$\partial_c f(x) = \partial_{c,0} f(x)$ the <u>c-subdifferential.</u> The elements of $\partial f(x)$ are called <u>c-subgradients</u> of f in x. There is the following characterization (cf. [19], [27], [1])

(70) $y \in \partial_c f(x) \Longleftrightarrow y \in E_c f(x)$ (i.e. $f(x) + f^*(y) = c(x,y)$)
             $\Longleftrightarrow f(x) - c(x,y) = \inf_{x'} (f(x') - c(x',y))$
             $(\Longleftrightarrow x \in \partial_c f(y)$, if f is c-convex).

If $\partial_{c,\epsilon} f(x) \neq \emptyset$ for all $0 < \epsilon \leq \epsilon_0$, then $f(x) = f^{**}(x)$.

   <u>Lemma</u> 19. Let $\Phi: E_1 \to E_2$, $\Phi(x) \in \partial_c f(x)$ for $x \in A$, then

(71) $c(y,\Phi(x)) + c(x,\Phi(y)) \leq c(x,\Phi(x)) + c(y,\Phi(y))$, $\forall x,y \in A$.

   <u>Proof.</u> Since $f(y) - f(x) \geq c(y,\Phi(x)) - c(x,\Phi(x))$ and $f(x) - f(y) \geq c(x,\Phi(y)) - c(y,\Phi(y))$, (71) follows by adding these inequalities.   □

   <u>Remark.</u>

a)   If $c(x,y) = -|x - y|^2$, $x,y \in \mathbb{R}^k$, then (71) is equivalent to the monotony of Φ,

(72) $\langle y - x, \Phi(y) - \Phi(x) \rangle \geq 0$.

   If $\Phi = \nabla f$, f continuous, differentiable, then from (35), this is necessary and sufficient for $\Phi(x) \in \partial_c f(x) = \partial f(x)$.

   If f,g,c are differentiable and $\Phi: \mathbb{R}^k \to \mathbb{R}^k$, then the condition that $\Phi(x) \in \partial_c f(x)$ implies that $h(y) := f(y) - f(x) - c(y,\Phi(x)) - c(x,\Phi(x)) \geq 0$ has a minimum in $y = x$ and, therefore,

(73) $\nabla f(x) = \partial_1 c(x,\Phi(x))$.

   If the differential form $\omega = \partial_1 c(x,\Phi(x)) \cdot dx$ is closed, we obtain

(74) $f(x) = c_1 + \int_{0 \to x} \partial_1 c(x,\Phi(x)) \cdot dx$.

   Similarly,

(75) $\nabla f^*(\Phi(x)) = \partial_2 c(x,\Phi(x))$

   and if Φ is invertible and $\partial_2 c(x,\Phi(x)) \cdot dx$ is closed, then

(76) $f^*(y) = c_2 + \int_{\Phi(0) \to y} \partial_2 c(\Phi^{-1}(u),u) \cdot du$.

   With the substitution $v = \Phi^{-1}(u)$, i.e. $du = \Phi'(v)dv$. We define $c_1 + c_2 = c(0,\Phi(0))$; then we obtain

(77) $f(x) + f^*(\Phi(x)) = c(0,\Phi(0)) + \int_{0 \to x} [\partial_1 c(u,\Phi(u)) \cdot u + \partial_2 c(u,\Phi(u))\Phi'(u)] \cdot du$
             $= c(0,\Phi(0)) + \int_{0 \to x} d(c(u,\Phi(u))) = c(x,\Phi(x))$.

Therefore, the condition that $\Phi(x)$ is the c-subgradient of a differenti function f, is equivalent to

(78) $f(x) + f^*(y) = f(x) - f(\Phi^{-1}(y)) + f(\Phi^{-1}(y)) + f^*(y) =$

   $c(\Phi^{-1}(y),y) + \int\limits_{\Phi^{-1}(y)\to x} \partial_1 c(u,\Phi(u)) \cdot du \geq c(x,y),$

equivalently, to the differential characterization

(79) $\int\limits_{\Phi^{-1}(y)\to x} [\partial_1 c(u,y) - \partial_1 c(u,\Phi(u))] \cdot du \leq 0, \ \forall x,y.$

(The case $c(x,y) = -|x-y|^\alpha$, $\alpha > 1$, has been considered in [96].)

As consequence of this discussion we obtain

   Theorem 20. If Φ is continuously differentiable, injective and if $\partial_i c(x,\Phi(x)) \cdot dx$, i = 1,2 is closed, then: $\Phi(x) \in \partial_c f(x)$, $\forall$ x, for a continuous differentiable function f if and only if (79) holds for all x,y.   □

3.2.   Generalizations of Young's Inequality

In this section we consider some generalizations of the Young-inequa-lity. Let $\Phi: [0,\infty) \to [0,\infty)$ be a Young-function i.e. Φ is right continuous, nondecreasing, $\Phi(0) = 0$ and $\Phi(x) \xrightarrow[x\to\infty]{} \infty$ and define the generalized inverse $\Phi^-(y) = \sup\{x: \Phi(x) \leq y\}$. The Young-inequality states the inequality:

(80) $xy \leq \int\limits_0^x \Phi(t)dt + \int\limits_0^y \Phi^-(s)ds$

for all x,y > 0 with equality if and only if $\Phi(x-) \leq y \leq \Phi(x)$ (cf. [6], [57], [17]).

Define for a measure generating function F on $[0,\infty)^2$ and corespond-ing measure m

(81) $h_1(x): = m\{(s,t); 0 \leq s \leq x, 0 \leq t \leq \Phi(s)\} = \int\limits_0^x (\int\limits_0^{\Phi(s)} dF(t|s))dF_1(s)$

   $h_2(y): = \int\limits_0^y (\int\limits_0^{\Phi^-(t)} dF(s|t))dF_2(t),$

$F(\cdot|\ )$, $F_1(\cdot)$ denote the conditional resp. marginal "distribution" functions.

   Theorem 21. For x,y > 0 we have:

(82) $F(x,y) + F(0,0) \leq (F(x,0) + h_1(x)) + (F(0,y) + h_2(y)).$

   Proof. Define A = $[0,x] \times [0,y]$, B = $\{(s,t); 0 \leq s \leq x, 0 \leq t \leq \Phi(s)\}$, C = $\{(s,t); 0 \leq t \leq y, 0 \leq s < \Phi^-(y)\}$. Then

(83) $A \subset B \cup C$ and $B \cap C = \emptyset.$

Therefore, m(A) = $F(x,y) - F(x,0) - F(0,y) + F(0,0) \leq m(B) + m(C) = h_1(x) + h_2(y).$

   □

Remark.

a)   The idea of the proof of Theorem 21 is due to Pales (1987) who noted that in the classical geometric proof one can use more general measures.

If $m = f \lambda_+^2$, then we obtain more explicitly:

(84) $$h_1(x) = \int_0^x (\int_0^{\Phi(s)} f(s,t)dt)ds = \int_0^x (\partial_1 F(s,\Phi(s)) - \partial_1 F(s,0))ds$$

$$= \int_0^x \partial_1 F(s,\Phi(s)) - F(x,0) + F(0,0)$$

and

$$h_2(y) = \int_0^y \partial_2 F(\Phi^-(t),t)dt - F(0,y) + F(0,0),$$

where the partial derivatives exist a.s. w.r.t. the Lebesgue measure. Therefore, from (82)

(85) $$F(x,y) \leq F(0,0) + \int_0^x \partial_1 F(s,\Phi(s))ds + \int_0^y \partial_2 F(\Phi^-(t),t)dt.$$

This inequality is due to Pales (1987) for $F \in C^2$ with $\partial_1 \partial_2 F(s,t) \geq 0$.

Example. Let for $\alpha > 1$, $F(x,y) = -|x - y|^\alpha$, $x,y \in R_+^1$, then $\partial_1 \partial_2 F(x,y) = \alpha(\alpha - 1)|x - y|^{\alpha - 2} \geq 0$. Therefore, by (86) we obtain the inequality

(86) $$|x - y|^\alpha \geq \alpha \int_0^x |s - \Phi(s)|^{\alpha - 1} sg(s - \Phi(s))ds$$

$$+ \alpha \int_0^y |t - \Phi^-(t)|^{\alpha - 1} sg(t - \Phi^-(t))dt,$$

where $sg(x) = \begin{cases} 1 & > \\ 0 & \text{if } x = 0 \\ -1 & < \end{cases}$ and Φ is a Young function. An analytical derivation of (86) has been given in [96]. A consequence of (86) and (62) is the wellknown fact that random variables (X,Y) with $\Phi(X-) \leq Y \leq \Phi(X)$ and Φ a Young function are optimal couplings w.r.t. the distance $c(x,y) = |x - y|^\alpha$.   □

We next derive an extension of (82) to the case of the whole real line.

Theorem 22. Let $\Phi: R^1 \to R^1$ be nondecreasing, right continuous. Let F be the generating function of a finite measure, $F(x,y) = P(X \leq x, Y \leq y)$, then

(87) $$F(x,y) \leq f(x) + g(y), \text{ where } f(x) = \int_{-\infty}^x P(Y \leq \Phi(s)|X = s)dP^X(s) \quad \text{and}$$

$$g(y) = \int_{-\infty}^x P(X < \Phi^-(s)|Y = s)dP(s).$$

   Proof. $F(x,y) = P(X \le x, Y \le \Phi(X) \wedge y) + P(X \le x, Y \le y, Y > \Phi(X)) \le$

$P(X \le x, Y \le \Phi(X)) + P(X < \Phi^-(Y), Y \le y) = \int_0^x P(Y \le \Phi(X)|X = s)dP^X(s) +$

$\int_0^x P(X > \Phi^-(Y)|Y = s)dP^Y(s) = f(x) + g(y).$ □

   In (87) we have equality, iff

(88) $\Phi(X-) \le Y \le \Phi(X)$ a.s.

(85), (87) imply optimal coupling results and Fréchet bounds for Δ-monotone
(resp. L-superadditive) functions (cf. (21)).

   An extension of the Young inequality to n-variables is the Oppenheim
inequality. Let $f_i: [0,\infty) \to [0,\infty)$ be Young-functions, $1 \le i \le n$, then:

(89) $\prod_{i=1}^{n} f_i(t_i) \le \sum_{i=1}^{n} \int_0^{t_i} (\prod_{j \ne i} f_j)df_i.$

This inequality was used in Gaffke and Rüschendorf (1981) to determine
$M_e(\varphi)$ for $\varphi(x) = \prod_{i=1}^{n} x_i$ and simple marginals $P_1,...,P_n$. For the literature cf.
Oppenheim (1927), Cooper (1927), and Dankert and König (1967).

   Consider the curve $y(t) = (f_1(t),...,f_n(t))$, $t \ge 0$ and the points
$P_i = (f_j(t_i))_{1 \le j \le n}$, $1 \le i \le n$, $A := (f_i(t_i))_{1 \le i \le n}$. Define, furthermore,

(90) $V_i := \{x \in \mathbf{R}_+; x_i \le f_i(t_i), x_j \le f_j(f_i^-(x_i)), j \ne i\}$ and

            $V := [0,A] = [0,f_1(t_1)] \times ... \times [0,f_n(t_n)].$

   Theorem 23. (Generalized Oppenheim Inequality)
Let m be a Radon measure on $\mathbf{R}_+^n$ and define $\varphi(t_1,...,t_n) := m([0,A])$,
$h_i(t_i) = m(V_i)$, $1 \le i \le n$, then:

(91) $\varphi(t_1,...,t_n) \le \sum_{i=1}^{n} h_i(t_i).$

   Proof. The proof follows from the inclusion $V = [0,A] \subset \bigcup_{i=1}^{n} V_i$
(cf. [65]).   □

   With $m = \boldsymbol{\lambda}^n$, the Lebesgue measure, we have from (91)

$\prod_{i=1}^{n} f_i(t_i) \le \sum_{i=1}^{n} \int_0^{f_i(t_i)} \prod_{j \ne i} f_j(f_i^-(s))ds = \sum_{i=1}^{n} \int_0^{t_i} (\prod_{j \ne i} f_j(u))df_i(u)$, the Oppenheim
inequality.   □

   For finite measures m we can extend (91) to $\mathbf{R}^n$.

   Theorem 24. Let $m = P^{(U_1,...,U_n)}$ be a finite measure on $\mathbf{R}^n$ with
generating function F and define: $\varphi(t_1,...,t_n) := F(f_1(t_1),...,f_n(t_n))$,

$h_i(t_i) := P(U_i \leq f_i(t_i), U_j \leq f_j \circ f_i^-(U_i), j \neq i) = \int_{-\infty}^{t_i} P(U_j \leq f_j(u), j \neq i \mid U_i = f_i(u)) dP^{f_i^-(U_i)}(u),$

then

(92) $\qquad \varphi(t_1,...,t_n) \leq \sum_{i=1}^{n} h_i(t_i).$

$\qquad$ <u>Proof.</u> $\varphi(t_1,...,t_n) \doteq P(U_i \leq f_i(t_i), 1 \leq i \leq n) = P(f_i^-(U_i) \leq t_i, 1 \leq i \leq n) \leq$

$\sum_{i=1}^{n} P(f_j^-(U_j) \leq t_j \wedge f_i^-(U_i), 1 \leq j \leq n) \leq \sum_{i=1}^{n} P(f_i^-(U_i) \leq t_i, f_j^-(U_j) \leq f_i^-(U_i), \forall j \neq i) \leq$

$\sum_{i=1}^{n} P(U_i \leq f_i(t_i), U_j \leq f_j \circ f_i^-(U_i), \forall j \neq i) = \sum_{i=1}^{n} \int_{-\infty}^{f_i(t_i)} P(U_j \leq f_j \circ f_i^-(s), \forall j \neq i \mid U_i = s) dP^{U_i}(s) =$

$\sum_{i=1}^{n} \int_{-\infty}^{t_i} P(U_j < f_j(u), j \neq i \mid U_i = f_i(u)) dP^{f_i^-(U)}(u).$ $\qquad \square$

4. <u>Some Statistical Applications and Problems</u>

4.1. <u>Marginal Sufficiency</u>

$\qquad$ Let $\mathfrak{P}$ be a dominated set of product measures on $(E,\mathfrak{A}) = \prod_{i=1}^{n}(E_i,\mathfrak{A}_i)$ and define $T: (E,\mathfrak{A}) \to (Y,\mathfrak{B})$ to be <u>marginally sufficient</u> for $\mathfrak{P}$, if for all $1 \leq i \leq n$ and $\varphi \in B(E,\mathfrak{A})$, $\varphi = \varphi(x_i)$, there exists $\tilde{\varphi} \in B(E,\sigma(T))$ with

(93) $\qquad \tilde{\varphi} = E_P(\varphi|T) \, [P], \ \forall P \in \mathfrak{P}.$

Huzurbazar proposed the following conjecture.

<u>Huzurbazar conjecture:</u>

(94) $\qquad$ Marginal sufficiency of T implies sufficiency

(i.e. (93) is true for any $\varphi = \varphi(x_1,...,x_n) \in B(E,\mathfrak{A})$).

$\qquad$ The first published proof of (94) was given by Sudakov (1979) in the case of equivalent measures (cf. also [53]). The idea of Sudakov's proof is related to some marginal problems. The idea is the following (cf. [105], p. 154 - 160). Let for $P = \otimes P_i$, $Q = \otimes Q_i \in \mathfrak{P}$, $f = \prod f_i$, $g = \prod g_i$ be densities w.r.t. a dominating measure $\mu = \otimes \mu_i$. Let $h_i = g_i / f_i$, $\varphi(x) := (\ell n \, h_1(x_1),...,\ell n \, h_n(x_n))$ and let P_y, Q_y denote the conditional distributions of P, Q given $T = y$. If T is marginally sufficient, then $\tilde{P}_y := P^{\varphi|T=y} = (P_y)^{\varphi}$ and $\tilde{Q}_y := Q^{\varphi|T=y} = (Q_y)^{\varphi}, y \in Y$, are probability measures on $\mathbb{R}^n$ with identical marginals. Using $\frac{dQ_y}{dP_y}(x) = \frac{dP^T}{dQ^T}(y) \prod h_i(x_i) = \frac{dP^T}{dQ^T}(y) \exp(\Sigma \, \ell n \, h_i(x_i))$ one concludes:

(95) $\qquad \frac{d\tilde{Q}_y}{d\tilde{P}_y}(z) = \frac{dP^T}{dQ^T}(y) \exp(\Sigma \, z_i).$

Let U, V be orthogonal, nonnegative measures with $P_y - Q_y = U - V$, then

1.   U,V have identical marginals,

2. $U(\Sigma x_k \le \ell) = 0$, $V(\Sigma x_k \ge \ell) = 0$ for some ℓ (namely $\ell = -\ell n \dfrac{dP^T}{dQ^T}(y)$),

3. $\int |x_k| dU < \infty$, $1 \le k \le n$.

To establish 3. is the most involved part of Sudakov's proof. It is easy to see that

(96)   1., 2., 3. implies that $U = V = 0$.

As consequence: $\tilde{P}_y = \tilde{Q}_y$ and, therefore, a standard argument from sufficiency theory implies that T is sufficient for $\{P,Q\}$. Since this holds for any pair P, Q, T is sufficient for $\mathfrak{P}$.

The following interesting example of Sudakov shows that the difficult moment condition 3. cannot be omitted.

Example. Let $\varphi: \mathbf{Z}^3 \to \mathbf{Z}^3$, $\varphi(x) = -x$ and define probability measures P,Q by

(97) $\dfrac{3}{4} P = \dfrac{1}{8} S\varepsilon_{(1,1,-1)} + \sum\limits_{k=2}^{\infty} \dfrac{1}{2^{k+2}} S\varepsilon_{(2^k-1,1-2^{k-1},1-2^{k-1})}$

         $Q := P^{\varphi}$,

where $S\varepsilon_{(1,1,-1)} = \varepsilon_{(1,1,-1)} + \varepsilon_{(1,-1,1)} + \varepsilon_{(-1,1,1)}$, ε_x the one point measure in x. Then the marginals of P, Q are identical and equal to

$\dfrac{1}{3} (\varepsilon_{-1} + \varepsilon_1) + \dfrac{4}{3} \sum\limits_{k=2}^{\infty} 2^{-(k+2)} (\varepsilon_{2^k-1} + \varepsilon_{(1-2^k)})$ and $P\{x \in \mathbf{Z}^3; \Sigma x_i = 1\} = 1$, while

$Q\{\Sigma x_i = -1\} = 1$.   □

Let more generally $\mathfrak{A}_1 \subset \ldots \subset \mathfrak{A}_n \subset \mathfrak{A}$ be an increasing sequence of σ-algebras and $P, Q \in M^1(E,\mathfrak{A})$. Let $Q \ll P$, $P_k = P/\mathfrak{A}_k$, $Q_k = Q/\mathfrak{A}_k$, $L_k = \dfrac{dQ_k}{dP_k}$ and $f_k := L_k / L_{k-1}$.

Theorem 25. (Generalized Huzurbazar conjecture, cf. [95])

If $T: (E,\mathfrak{A}) \to (Y,\mathfrak{B})$ is partially sufficient for $\sigma(f_k)$, $1 \le k \le n$, then T is sufficient for $\{P_n, Q_n\}$.

The proof uses the following two lemmas:

Lemma 1. (cf. [78], Prop. 6)

If $P_i \in M^1(\mathbf{R}^1)$, $1 \le i \le n$, $P,Q \in M(P_1,\ldots,P_n)$, then:

(98) $P \le_{st} Q$ implies $P = Q$.   □

Lemma 2. (cf. Simons (1980))

For any sub-σ-algebra $\mathcal{B} \subset \mathcal{U}$ and $P, Q \in M^1(E, \mathcal{U})$ the conditional distributions of $L = (L_1, ..., L_n)$ w.r.t. $\mathcal{B}$ satisfy:

(99) $P^{L | \mathcal{B}} \leq_{st} Q^{L | \mathcal{B}}$.   □

If T is partially sufficient for $\sigma(f_k)$, $1 \leq k \leq n$, then $P_{\mathcal{B}}^{(f_1, ..., f_n)}, Q_{\mathcal{B}}^{(f_1, ..., f_n)}$ have the same marginals, where $P_{\mathcal{B}}$, $Q_{\mathcal{B}}$ are the conditional distributions. Then by a generalization of the a.s. representation theorem of stochastic orders to Markov kernels, one obtains versions $X_{\mathcal{B}}$, $Y_{\mathcal{B}}$ of the distribution in (99) such that $X_{\mathcal{B}} \leq Y_{\mathcal{B}}$ a.s. From these versions one can construct versions $\tilde{X}_{\mathcal{B}}$, $\tilde{Y}_{\mathcal{B}}$ of the distributions of $P_{\mathcal{B}}^{(f_1, ..., f_n)}, Q_{\mathcal{B}}^{(f_1, ..., f_n)}$ such that $\tilde{X}_{\mathcal{B}} \leq \tilde{Y}_{\mathcal{B}}$ a.s. Therefore, from (98) one obtains $P_{\mathcal{B}}^{(f_1, ..., f_n)} = Q_{\mathcal{B}}^{(f_1, ..., f_n)}$. This implies that $\mathcal{B}$ is sufficient, $\mathcal{B} = \sigma(T)$.

4.2. Optimal Combination of Tests of Marginals

Let P_i, $Q_i \in M^1(E_i, \mathcal{U}_i)$, $1 \leq i \leq n$, and consider the testproblem with hypothesis $\Theta_o = M(P_1, ..., P_n)$ and alternative $\Theta_1 = M(Q_1, ..., Q_n)$. In a practical problem this means e.g. that one measures n components and has for each components the simple alternatives $\{P_i\}$, $\{Q_i\}$ but does not know anything about the dependence structure of the measurements. The question then is the following: Is it possible to achieve a better test Θ_o, Θ_1 then to take the test for that component which allows for a certain test level α the highest power? What is the optimal combination of the marginal tests?

The answer to this problem was given in [85] w.r.t. the maximin criterion. We consider the tests of level α

(100) $\Phi_\alpha(\Theta_o) = \{\varphi \in \Phi : E_P \varphi \leq \alpha, \ \forall P \in M(P_1, ..., P_n) = \Theta_o\}$

and the maximinrisk

(101) $\beta(\alpha, \Theta_o, \Theta_1) = \sup_{\varphi \in \Phi_\alpha(\Theta_o)} \inf_{P \in \Theta_1} E_P \varphi$.

Let for two finite measures P,Q on $(E, \mathcal{U})$

(102) $d_v(P, Q) = \sup \{P(A) - Q(A); \ A \in \mathcal{U}\}$

             $d_v(\mathcal{P}, \mathcal{Q}) = \inf \{d_v(P, Q); \ P \in \mathcal{P}, \ Q \in \mathcal{Q}\}$

for subsets $\mathcal{P}, \mathcal{Q} \subset M(E, \mathcal{U})$ and define

(103) $h_\alpha(x) := \alpha x + \max_{1 \leq i \leq n} d_v(Q_i, x P_i), \ x \geq 0$.

Theorem 26. (cf. [85])

Let $\alpha \in (0,1)$ and let x be a minimum point of h_α, then:

a) $\beta(\alpha,\Theta_0,\Theta_1) = h_\alpha(x^*)$.

b)   If $P \in \Theta_0$, $Q \in \Theta_1$ satisfy

(104) $d_v(Q, x^* P) = d_v(\Theta_1, x^* \Theta_0)$,

then there exists a LQ-test φ^* for $(\{P\},\{Q\})$ with critical value x^* such that φ^* is a maximin level α-test, i.e.

(105) $\varphi^* \in \Phi_\alpha(\Theta_0)$ and $\inf_{P \in \Theta_1} E_P \varphi^* = \beta(\alpha,\Theta_0,\Theta_1)$.   □

The proof uses the following lemma.

Lemma 3. $\forall x \geq 0$ holds:

(106) $d_v(M(Q_1,...,Q_n), x M(P_1,...,P_n)) = \max_{1 \leq i \leq n} d_v(Q_i, x P_i)$.   □

Minimal pairs can explicitly be determined. Furthermore, the proof uses a characterization of maximintests given by Baumann (1968).

One can not improve the best test of single marginals if e.g.

(107) $d_v(Q_1, x P_1) = \max_j d_v(Q_j, x P_j)$, $\forall x \geq 0$.

But in other cases one obtains a considerable improvement. Some related results with additional restrictions on the hypotheses have been discussed in [93].

An alternative interpretation of Theorem 26 is in terms of robustness. If $M(P_1,...,P_n)$ is considered as a neighbourhood of $P_1 \otimes ... \otimes P_n$, then for a test φ, $M_\alpha(\varphi)$ is its robust level and Theorem 26 constructs an optimal robust test.

4.3.   Optimal Estimators in Marginal Models

We consider the construction of minimum variance unbiased estimators (MVU) in the model $\mathfrak{P} = M(P_1,...,P_k)$ for certain functions $g : \mathfrak{P} \to \mathbb{R}^1$. The general question is the following: How can one use the knowledge of the marginals in order to construct better estimators than those in the model without this knowledge?

Let D_o be the set of all unbiased estimators of zero, let $\bar{P} := \bigotimes_{i=1}^{k} P_i$ and let D_g denote the unbiased estimators of g.

Theorem 27. (cf. [89])

a) $D_o = F_1^\perp = \{\sum_{i=1}^{k} f_i(x_i); \ f_i \in \mathcal{B}^1(P_i), \ \int f_i dP_i = 0, \ 1 \leq i \leq k\}$.

b) If $P \in \mathcal{P}$ and $d \in D_g \cap L^2(P)$, then

(108) $d^*: = d - E_p(d|\overline{F}_2^P)$

is MVU for g in P, where $\overline{F}_2^P$ denotes the closure of $F_2 = \{\sum f_i(x_i);$ $f_i \in \mathcal{B}^2(P_i), \ \int f_i dP_i = 0\}$ in $\mathcal{B}^2(P)$.

c)   If $d \in D_g$, then

(109) $d^*: = d - \sum_{i=1}^{k} \int d \ d \bigotimes_{j \neq i} P_j + k \int d \ d\overline{P}$

is MVU for g in $\overline{P}$.

The projections occuring in Theorem 27 can be calculated in some cases while in the general case an approximative solution based on the alternating projection theorem is known (cf. [84]).

In the case of n independent observations the underlying model is $\mathcal{P}^n = \{P^n; \ P \in \mathcal{P}\}$ and the corresponding optimal estimator is given by

(110) $d_n^*(x_1,...,x_n) = \frac{1}{n} \sum_{i=1}^{n} d^*(x_i)$.

An estimator sequence for a differentiable functional g, which is asymptotically optimal on the whole model or a subset $\mathcal{P}_o \in \mathcal{P}$, should have the stochastic expansion

(111) $\sqrt{n} \ (d_n^*(x_1,...,x_n) - g(P)) = \frac{1}{\sqrt{n}} \sum_{i=1}^{n} g_P(x_i) + o_{pn}(1), \ P \in \mathcal{P}_o$,

where g_P lies in the tangent cone $T(P,\mathcal{P})$, the set of all derivatives (tangent vectors) of L^2-differentiable path's in $\mathcal{P}$ through P. $T(P,\mathcal{P})$ can be shown to be identical to

(112) $T(P,\mathcal{P}) = (F_2)^{\perp P}$,

the orthogonal complement of F_2 in $L^2(P)$ (cf. [90], [92]). The stochastic expansion in (111) implies that g_P is a gradient of g and since $g_P \in T(P,\mathcal{P})$, it is the canonical gradient.

In a recent paper Bickel, Ritov and Wellner (1988) succeeded to construct an estimator sequence with this property on the subset $\mathcal{P}_\alpha \subset \mathcal{P}$, $\alpha > 0$, k = 2, consisting of 'positive dependent' measures P with

(113) $P(A \times B) \geq \alpha P_1(A) P_2(B), \ \forall A, B$.

References

[1] Balder, S. J.: An extension of duality-stability relations to non-convex optimization problems. SIAM J. Contr. Opt. 15 (1977), 320 - 343

[2] Baumann, V.: Eine parameterfreie Theorie der ungünstigsten Verteilungen. Z. Wahrscheinlichkeitstheorie verw. Geb. 11 (1968), 40 - 60

[3] Bertino, S.: Su di una sottoclasse della classe di Fréchet. Statistica 28 (1968), 511 - 542

[4] Bertino, S.: Sulla distanza tra distribuzioni. Pubbl. Ist. Calc. Prob. Univ. Roma, n. 82 (1968)

[5] Bickel, P. J., Ritov, Y., and Wellner, J. A.: Efficient estimation of a probability measure P with known marginal distributions. Preprint, 1988

[6] Boas, R. P., and Marcus, M. B.: Inverse functions and integration by parts. Amer. Math. Monthly 81 (1974), 760 - 761

[7] Brown, J. R.: Approximation theorems for Markov operators. Pacific J. Math. 16 (1966), 13 - 23

[8] Cambanis, S., Simons, G., Stout, W.: Inequalities for $Ek(X,Y)$ when the marginals are fixed. Z. Wahrscheinlichkeitstheorie 36 (1976), 285 - 294

[9] Cooper, R.: Notes on certain inequalities. J. London Math. Soc. 2 (1927), 17 - 21

[10] Csiszar, I.: I-divergence geometry of probability distributions and minimization problems. Ann. Prob. 3 (1975), 146 - 158

[11] Dall'Aglio, G.: Sugli estremi dei momenti delle funzioni di ripartizione dopia. Ann. Scuola Normale Superiore Di Pisa, ser. Cl. Sci., 3.1 (1956), 33 - 74

[12] Dall'Aglio, G.: Sulla compatibilita delle funzioni di ripartizione doppia. Rendiconti di Math. 18 (1959), 385 - 413

[13] Dall'Aglio, G.: Les fonctions extrêmes de la classe de Frêchet à 3 dimensions. Publ. Inst. Stat. Univ. Paris, IX (1960), 175 - 188

[14] Dall'Aglio, G.: Fréchet classes and compatibility of distribution functions. Symp. Math. 9 (1972), 131 - 150

[15] Dankert, G., and König, H.: Über die Höldersche Ungleichung in Orlicz-Räumen. Arch. Math. 118 (1967), 61 - 75

[16] Darroch, J. N., Lauritzen, S. L., and Speed, T. P.: Markov fields and log-linear interaction models for contingency tables. Ann. Statist. 8 (1980), 522 - 539

[17] Diaz, J. B., and Metcalf, F. T.: n analytic proof of Young's inequality. Amer. Math. Monthly 77 (1970), 603 - 609

[18] Dobrushin, R. L.: Prescribing a system of random variables by conditional distributions. Theory Prob. Appl. 15 (1970), 458 - 486

[19] Dolecki, S., and Kurcyusz, St.: On Φ-convexity in extremal problems. SIAM J. Control. Optim. 6 (1978), 277 - 300

[20] Douglas, R. G.: On extremal measures and subspace density. Michigan Math. J. 11 (1964), 243 - 246

[21] Dowson, C. D., and Landau, B. U.: The Fréchet distance between multivariate normal distributions. J. Multivar. Anal. 12 (1982), 450 - 455

[22] Dudley, R. M.: Distances of probability measures and random varaibles. Ann. Math. Statist. 39 (1968), 1563 - 1572

[23] Dudley, R. M.: Probability and Metrics. Aarhus Univ., Aarhus, 1976

[24] Dudley, R. M.: Real Analysis and Probability. Wadsworth and Brooks/Cole, 1989

[25] Dunford, N., and Schwartz, J. T.: Linear Operators, Part I. Interscience Publishers, 1967

[26] Edwards, D. A.: On the existence of probability measures with given marginals. Ann. Inst. Fourier 28 (1978), 53 - 78

[27] Elster, K. H., and Nehse, R.: Zur Theorie der Polarfunktionale. Math. Operationsf. u. Statist. 5 (1974), 3 - 21

[28] Fernique, X.: Sur le théorème de Kantorovitch-Rubinstein dans les espaces polonais. Lecture Notes in Mathematics 850, pp. 6 - 10. Springer, 1981

[29] Frank, M. J., Nelsen, R. B., and Schweizer, B.: Best possible bounds for the distribution of a sum - a problem of Kolmogorov. Prob. Theory Rel. Fields 74 (1987), 199 - 211

[30] Fréchet, M.: Sur les tableaux de corrélation dont les marges sont donnees. Annales de l'Universite de Lyon, Sciences 4 (1951), 13 - 84

[31] Fréchet, M.: Les Tableaux de corrélation dont les marges et des bornes sont donnees. Annales Univ. Lyon, Sciences 20 (1957), 13 - 31

[32] Fréchet, M.: Sur la distance de deux lois de probabilité. Publ. Inst. Stat. Univ. Paris 6 (1957), 185 - 198

[33] Fréchet, M.: Sur les tableaux de corrélation dont les marges et des bornes sont donnees. Revue Inst. Int. de Statistique 28 (1960), 10 - 32

[34] Gaffke, N., and Rüschendorf, L.: On a class of extremal problems in statistics. Math. Operationsforschung Stat., Ser. Optimization 12 (1981), 123 - 135

[35] Gaffke, N., and Rüschendorf, L.: On the existence of probability measures with given marginals. Statistics & Decisions 2 (1984), 163 - 174

[36] Gini, C.: Di una misura della dissomiglianza tra due gruppi di quantità e delle sue applicazioni allo studio delle relazioni statistiche. Att. R. Ist. Veneto Sc. Lettere, Art 74 (1914), 185 - 213

[37] Givens, C. R., and Shortt, R. M.: A class of Wasserstein metrics for probability distributions. Manuscript, Dept. Math. Comp. Sci. Michigan Tech. Univ., Houghton, MI, 1984

[38] Hammersley, I. M., and Handscomb, D. C.: Monte Carlo Methods. Meth London, 1964

[39] Hansel, G., and Troallic, J. P.: Mesures marginales et théorème de Ford-Fulkerson. Z. W.-theorie verw. Gebiete 43 (1978), 245 - 251

[40] Hoeffding, W.: Masstabinvariante Korrelationstheorie. Sem. Math. Inst. Univ. Berlin 5 (1950), 181 - 233

[41] Huang, J. S., and Kotz, S.: Correlation structure in iterated Farlie-Gumbel-Morgenstern distributions. Biometrika 71 (1984),

[42] Ivanov, E. H., and Nehse, R.: Relations between generalized concepts of convexity and conjugacy. Math. Operationsforsch. Statist., Ser. Opt. 13 (1982), 9 - 18

[43] Johnson, N. L., and Kotz, S.: On some generalized Farlie-Gumbel-Morgenstern distributions. Comm. Statistics 4 (1975), 415 - 427

[44] Kamae, T., Krengel, U., and O'Brien, G. L.: Stochastic inequalities on partially ordered spaces. Ann. Prob. 5 (1977), 899 - 912

[45] Kantorovic, L. V., and Rubinstein, G. S.: On a space of completely additive functions (in Russian). Vestnik Leningrad Univ. 13 (1958), 52 - 59

[46] Kellerer, H. G.: Verteilungsfunktionen mit gegebenen Marginalvertei-lungen. Z. Wahrsch. 3 (1964), 247 - 270

[47] Kellerer, H. G.: Duality theorems for marginal problems. z. Wahrsch. 67 (1984), 399 - 432

[48] Kellerer, H. G.: Measure theoretic versions of linear programming. Preprint, 1987

[49] Kimeldorf, G., and Sampson, A.: Monotone dependence. Ann. Statist. 6 (1978), 895 - 903

[50] Klein Haneveld, W. K.: Robustness against PERT: An application of duality and distributions with known marginals. Preprint, 1984

[51] Knott, M., and Smith, C. S.: On the optimal mapping of distributions. J. Optim. Th. Appl. 43 (1984), 39 - 49

[52] Kotz, S., and Johnson, N.: Propriétés de dépendence des distributions iterees generalisees a deux variables Farlie-Gumbel-Morgenstern. C. R. Acad. Sc. Paris 285 (1977), 277 - 280

[53] Kudo, H.: On marginal sufficiency. Statistics & Decisions 4 (1986), 301 - 320

[54] Lai, T. L., and Robbins, M.: Maximally dependent random variables. Proc. Nat. Acad. Sci. USA, 73 (1976), 286 - 288

[55] Lancaster, H. O.: Correlation and complete dependence of random variables. Ann. Math. Statist. 34 (1963), 1315 - 1321

[56] Lauritzen, S. L., and Wermuth, N.: Graphical models for associations between variables. Ann. Statist. 17 (1989), 31 - 57

[57] Lembcke, J.: Gemeinsame Urbilder endlich additiver Inhalte. Math. Ann. 198 (1972), 239 - 258

[58] Levin, V. L., and Malyutin, A. A.: The mass transfer problem with discontinuous cost function and a mass setting for the problem of duality of convex extremum problems. Uspekhi Mat. Nauk 34 (1979), 3 - 68 (in Russian)

[59] Losonczi, L.: Inequalities of Young-type. Monatsh. Math. 97 (1984), 125 - 132

[60] Luschgy, H., and Thomsen, W.: Extreme points in the Hahn-Banach-Kantorovic setting. Pacific J. Math. 105 (1983), 387 - 398

[61] Makarov, G. D.: Estimates for the distribution function of a sum of two random variables when the marginal distributions are fixed. Theory Prob. Appl. 26 (1981), 803 - 806

[62] Mardia, K. V.: Families of Bivariate Distributions. Hafner, Darim, 1970

[63] Mosler, K. C.: Entscheidungsregeln bei Risiko: Multivariate stochastische Dominanz. Lecture Notes in Economics and Math. Systems 204. Springer, 1982

[64] Olkin, I., and Pukelsheim, F.: The distance between two random vectors with given dispersion matrices. Linear Algebra and Appl. 48 (1982), 257 - 263

[65] Oppenheim, A.: Note on Cooper's generalization of Young's inequality. J. London Math. Soc. 2 (1927), 21 - 23

[66] Pales, Z.: A Generalization of Young's inequality. In: Inequalities V. Ed. E. Walter (1987)

[67] Plackett, R. L.: A class of bivariate distributions. J. Am. Stat. Assoc. 60 (1965), 516 - 522

[68] Rachev, S. T.: On a problem of Dudley. Soviet Math. Dokl. 29 (1984), 162 - 164

[69] Rachev, S. T.: The Monge Kantorovich mass transference problem and its stochastic applications. Theroy Prob. Appl. 24 (1985), 647 - 671

[70] Rachev, S. T., and Shortt, R. M.: Duality theorems for Kantorovic-Rubinstein and Wasserstein functions. Preprint, 1989

[71] Rachev, S. T., and Rüschendorf, L.: A transformation property of minimal metrics. To appear in: Theory Prob. Appl., 1989

[72] Rachev, S. T., Rüschendorf, L., and Schief, A.: On the construction of almost surely convergent random variables. Preprint: Angew. Math. und Informatik 10 (1988)

[73] Rachev, S. T., and Rüschendorf, L.: A counterexample to a.s. constructions. Stat. Prob. Letters 9 (1990), 307 - 309

[74] Rockafellar, R. T.: Convex Analysis. Princeton Univ. Press, 1970

[75] Rüschendorf, L.: Vergleich von Zufallsvariablen bzgl. integralinduzierter Halbordnungen. Habilitationsschrift, Aachen, 1979

[76] Rüschendorf, L.: Inequalities for the expectation of Δ-monotone functions. Z. Wahrsch. 54 (1980), 341 - 354

[77] Rüschendorf, L.: Ordering of distributions and rearrangement of functions. Ann. Probab. 9 (1980), 276 - 283

[78] Rüschendorf, L.: Stochastically ordered distributions and monotonicity of the OC of an SPRT. Math. Operationsf., Statistics 12 (1981), 327 - 338

[79] Rüschendorf, L.: Sharpness of Frechet bounds. Z. Wahrsch. 57 (1981), 293 - 302

[80] Rüschendorf, L.: Random variables with maximum sums. Adv. Appl. Prob. 14 (1982), 623 - 632

[81] Rüschendorf, L.: Solution of a statistical optimization problem by rearrangement methods. Metrika 30 (1983), 55 - 62

[82] Rüschendorf, L.: On the multidimensional assignment problem. Methods of OR 47 (1983), 107 - 113

[83] Rüschendorf, L.: On the minimum discrimination information theorem. Statistics and Decisions, Suppl. Issue No. 1 (1984), 263 - 283

[84] Rüschendorf, L.: Projections and iterative procedures. In: Proceedings Sixth Intern. Symp. Mult. Anal., Pittsburgh. Ed. P. R. Krishnaiah (1985), 485 - 593

[85] Rüschendorf, L.: Robust tests against dependence. Prob. Math. Statistics 6 (1985), 1 - 10

[86] Rüschendorf, L.: The Wasserstein distance and approximation theorems. Z. Wahrsch. 70 (1985), 117 - 129

[87] Rüschendorf, L.: Construction of multivariate distributions with given marginals. Ann. Inst. Stat. Math. 37 (1985), 225 - 233

[88] Rüschendorf, L.: Monotonicity and unbiasedness of tests via a.s. constructions. Statistics 17 (1986), 221 - 230

[89] Rüschendorf, L.: Unbiased estimation in nonparametric classes of distributions. Statistics and Decisions 5 (1987), 89 - 104

[90] Rüschendorf, L.: Unbiased estimation and local structure. Proceedings 5th Pannonian Symposium in Visegrad (1985), 295 - 306

[91] Rüschendorf, L.: Projections of probability measures. Statistics 18 (1987), 123 - 129

[92] Rüschendorf, L.: Unbiased estimation of von Mises functionals. Statist. Prob. Letters 5 (1987), 287 - 292

[93] Rüschendorf, L.: Maximintests for neighbourhoods caused by dependence. In: Proceedings 1st World Congress of the Bernoulli Soc., Tashkent, 1986

[94] Rüschendorf, L., and Rachev, S. T.: A characterization of random variables with minimum L^2-distance. J. Mult. Analysis 32 (1990), 48 - 54

[95] Rüschendorf, L.: Conditional stochastic order and partial sufficiency. To appear in: Adv. Appl. Prob., 1989

[96] Rüschendorf, L.: Bounds for distributions with multivariate marginals. To appear in: Proceedings: Stochastic Order and Decisions under Risk. Ed. K. Mosler and M. Scarsini, 1989

[97] Salvemini, T.: Nuovi procedimenti per il calcolo degli indici di dissomiglianza e di connessione. Statistica (1949), 3 - 26

[98] Scarsini, M.: Lower bounds for the distribution function of a k-dimensional n-extendible exchangeable process. Statist. Prob. Letters 3 (1985), 57 - 62

[99] Scarsini, M.: Copulae of probability measures on product spaces. To appear in: J. Mult. Anal., 1989

[100] Shortt, R. M.: Combinatorial methods in the study of marginal problems over separable spaces. J. Math. Anal. 97 (1983), 462 - 479

[101] Simons, G.: Extensions of the stochastic ordering property of likelihood ratios. Ann. Statist. 8 (1980), 833 - 839

[102] Snijders, T. A. B.: Antithetic variates for Monte Carlo estimation of probabilities. Statistics Neerlandica 38 (1984), 1 - 19

[103] Stoyan, D.: Comparison Methods for Queues and other Stochastic Models. Wiley, 1983

[104] Strassen, V.: The existence of probability measures with given marginals. Ann. Math. Statist. 36 (1965), 423 - 439

[105] Sudakov, V. N.: Geometric problems in the theory of infinite dimensional probability distributions. Proc. Steklov Institute 141 (1979), 1 - 178

[106] Szulga, A.: On the Wasserstein metric. In: Transactions of the 8th Prague Conf. on Inform. Theory, Statist. Decision Funct., and Random Processes, Prague, 1978, v. B. Akademia, Praha, pp. 267 - 273

[107] Szulga, A.: On minimal metrics in the space of random variables. Theory Prob. Appl. 27 (1982), 424 - 430

[108] Tchen, A. H.: Inequalities for distributions with given marginals. Ann. Probab. 8 (1980), 814 - 827

[109] Vallender, S. S.: Calculation of the Wasserstein distance between probability distributions on the line. Theory Prob. Appl. 18 (1973), 784 - 786

[110] Vorobev, N. N.: Consistent families of measures and their extensions. Theory Prob. Appl. 7 (1962), 147 - 163

[111] Warmuth, W.: Marginal Fréchet-bounds for multidimensional distribution functions. Statistics 19 (1988), 283 - 294

[112] Whitt, W.: Bivariate distributions with given marginals. Ann. Statist. 4 (1976), 1280 - 1289

EXTREMAL SOLUTIONS IN THE MARGINAL PROBLEM

Viktor Benes
Czech Tecnical University, Dept. of Mathematics
Karlovo nam 13
12135 Praha 2
Czechoslovakia

Josef Stepan
Charles University, Dept of Statistics
Sokolovska 83
18600 Praha 8
Czechoslovakia

ABSTRACT. This is a review paper collecting both older and recent progress in the field of twodimensional marginal problem concerning the extreme points of the set of solutions. Some new results in the characterization of extremal solutions are presented in Section 5.

In Section 1 the problem is formulated as a special case of the moment problem in probability theory. Section 2 presents the review of the discrete case starting with Birkhoff's permutation matrices, over Letac's set of marginal uniqueness to the recent combinatorial approaches. Section 3 reviews the continuous case including the functional-analytic characterization of Douglas, measure theoretic techniques based on rectangles and methods studying directly the supports of extremal solutions on graphs.

The approach of the authors in Section 4 leads to the characterization of the sets of marginal uniqueness on two graphs by means of invariant measures. Using the Choquet theory and previous result by Stepan a new necessary and sufficient condition for an extremal solution is derived in Section 5, which is more geometrically objective. Section 6 yields only a few results concerning the multidimensional case.

1. Introduction.

If C is a bounded convex set in a locally convex vector topological space E, we shall denote its set of extreme points by exC. There are many reasons to investigate the set exC. Perhaps the most important one is that in many situations the extreme points are "rich enough" to generate points in C in the sense of Krein-Milman and Choquet theorem. In this case it is

$$\sup_{x \in C} g(x) = \sup_{x \in exC} g(x) \tag{1}$$

189

where g is a convex functions on C. This relation may provide a substancial reduction for some optimization problems.

Situation (1) is really frequent in probability theory. An example is provided by the following theorem (Bican and Stepan (1985), Winkler and Weizsacker (1980)):

Let X be a Polish space and consider a convex set of Borel probability measures on X

$$C = \{P \in H; \ Pt_j^{-1} = P, \ \int_X g_j \ dP = 0, \ j \in N\}$$

where H is a convex weakly closed set of probability measures, and t_j : $X \to X$, $g_j : X \to R$ are Borel measurable maps. Then for every $P \in C$ there is a probability measure m supported by exC which represents P in the sense

$$P(B) = \int_{exC} Q(B) \ m(dQ)$$

The measure m is defined on the σ-algebra of subsets of exC which makes $Q \to Q(B)$, B Borel subset of X, measurable. Moreover, (1) is valid for any $g: C \to R$ that is bounded convex and continuous with respect to the weak topology in C, while the existence of $\max_C g$ implies the existence of $\max_{exC} g$.

The marginal problem, or better to say the set C of its solutions, is of the type presented by the preceeding theorem. We shall treat here this very simple case.

Consider probability spaces $(X_i, \mathcal{B}_i, P_i)$, i=1,2 and denote

$$(X, \mathcal{B}) = (X_1 \times X_2, \mathcal{B}_1 \times \mathcal{B}_2)$$

and put

$$C = \{P \text{ probability measure on } \mathcal{B}; \ P^i = P_i , i = 1,2\} \tag{2}$$

where we have denoted the marginal probability measures of P by P^i , i.e. $P^i(B) = P(B \times X_i)$, $B \in \mathcal{B}_i$.

To realize the program suggested by (1) it is necessary to characterize the elements in exC or more precisely to find

> *sufficient and necessary conditions for a set D to support* $\qquad$ (3)
> *a probability measure $P \in exC$.*

This problem has not yet been solved generally and it is our aim to survey the state of research in this field.

Although we do not intend to discuss applications of the marginal problem generally, recall for example that if X_1 and X_2 are finite sets with cardinality n, then C may be thought as the set of feasible programs of the transportation problem in a linear programming theory and the set exC as the set of support programs. Applications in the continuous case are discussed by Dall'Aglio (1972) and many others.

In Section 2 we start the review with the discrete case of X_i finite or countably infinite. Some authors solve the general problem (2), others deal with doubly stochastic matrices which correspond to measures having each row and column sums of atoms equal to one. In the countably infinite case the uniform marginal measures of doubly stochastic matrices are infinite.

Other possibilities are to define

$$C_1 = \{P \text{ on } \mathfrak{B}, P^1 = P_1\} \qquad \text{or} \qquad C_2 = \{P \text{ on } \mathfrak{B}, P^i \leq P_i, \ i=1,2\}.$$

Then we speak about the sets C_i of stochastic, doubly substochastic matrices (measures), respectively.

In Section 3 the continuous spaces X_1 are considered. The problem is again formulated either for a general set C in (2) or for the special case $X_1 = X_2 = I$ (the unit real interval $< 0,1>$), P_i Lebesgue measures on Borel σ-algebras $\mathfrak{B}_i$ of I. In the latter case C is the set of the so called doubly stochastic measures. In both Section 2 and 3 the aim is not to present an exhaustive survey. We apologize with the authors not mentioned, we want only remember briefly the main ideas of the theory.

The general approach to the marginal problem as introduced in Benes and Stepan (1987) is presented in Section 4. Section 5 yields a necessary and sufficient condition for a measure to be extreme in C using the results in Stepan (1977), (1979). In Section 6 some generalizations of the theory towards the product of more than two spaces are reviewed.

2. The extreme points in the discrete marginal problem.

The survey is started by the theorem of Birkhoff (1946) which says that the space C_n of all doubly stochastic n x n matrices is equal to the convex hull of the set of permutation matrices, i.e. matrices with exactly one non-zero element in each column and row.

Remark that the representation of a doubly stochastic (d.s.) matrix does not require all n! permutation matrices. In fact, C_n lies in a linear variety of dimension $(n-1)^2$ in $\mathbb{R}^{n^2}$. By Caratheodory's theorem any d.s. matrix belongs to the convex hull of at most $(n-1)^2 + 1$ permutation matrices. This result cannot be improved, see Hammersley (1961).

One consequence of Birkhoff's theorem is that any function of a matrix variable defined and convex on C_n assumes its maximum for a permutation matrix. Inequalities in matrix theory were obtained in this way by Marcus (1960). There exist analogues of Birkhoff's theorem for stochastic and doubly substochastic matrices (Mirsky, 1959).

The results just mentioned coincide in fact with Krein-Milman property leading to a possibility to use optimization procedure (1). Indeed, the set of permutation matrices coincides with the set of extreme points $\text{ex}C_n$; see for example Hoffmann and Wielandt (1953).

In his book "Lattice Theory" Birkhoff proposed the problem of extending his theorem to the set C of denumerably infinite d.s. matrices. This question has acquired some celebrity as "Birkhoff's problem 111". Let Y be a vector space of infinite matrices such that $C \subset Y$ and denote with C_e the set of infinite permunation matrices. It is then required to prove that, for a suitable choice of Y and suitable topology on Y compatible with its linear structure, the set C is the closed convex hull of C_e.

The solution was given independently by Rattray and Peck (1955) and Kendall (1960). Kendall showed again that each extreme point in C is a permutation matrix, i.e. C_e=exC. The difference between the finite and infinite case is that Birkhoff's theorem follows immediately using the characterization of extreme points, which is not true in the infinite case.

Isbell (1962) proved (using the axiom of choice) that an infinite d.s. matrix can be expressed as a finite convex combination of extreme points if and only if elements take only finitely many distinct values.

More references as well as some open problems to the topic of d.s. matrices are mentioned in Mirsky's (1963) review paper.

The general set C in (2) was investigated in the discrete case by some authors later. Letac (1966) deals with countably infinite X_i with discrete topology. A sequence of points x∈X,

$$x=(x_i, y_i)_{i=1}^{2n} \quad \text{for which}$$

$$x_{2k-1}=x_{2k}, 1 \leq k \leq n, \qquad y_{2k} = y_{2k+1}, 1 \leq k \leq \text{n-1}, \qquad y_{2n}=y_1$$

is called a cycle. F∈𝔅 is a set of uniqueness if any probability measure μ with support supp$\mu \subset$ F is determined by its marginals. The three following assertions are equivalent:

a) $\mu \in$ exC
b) suppμ is a set of uniqueness   (4)
c) suppμ does not contain any cycle.

Given a set F∈𝔅 denote

$$C_F = \{P \in C; \text{supp}P \subset F\} \tag{5}$$

If F is a set of uniqueness then cardC_F is either 0 or 1. Conditions are given for the existence problem, i.e. for C_F to be non-empty.

A more detailed discussion of the discrete finite case is presented in Diego and Germani (1972). The sets of uniqueness are there called trees. Maximal and disconnected trees are characterized, an algorithm for the computation of an extremal measure given its support tree and marginals is proposed. Among many results we choose the following: F is a support of an extremal measure if and only if F is a minimal set among those F_i satisfying

$$A \subset X_1, \; B \subset X_2 \text{ nonempty}, \; (A \times B) \cap F_i = \emptyset \Rightarrow P_1(A) + P_2(B) \leq 1.$$

Coming back to countably infinite X_i's, Denny(1980) yields another equivalent condition to (4) which will be discussed in more detail in Section 4.

Mukerjee(1985), discussing Letac's and Denny's results, emphasizes the combinatorial nature of the characterization of extreme measures. His proofs are simpler and geometrically intuitive. A complete description of the support of extreme measures is given. A strengthened version of the Douglas-Lindenstrauss theorem (see Section 3) is proved for the discrete case as follows: $\mu \in$ exC if and only if for every real funcion h on X there exist real functions p on X_1, q on X_2 such that

$$h(x,y) = p(x) + q(y) \quad \mu\text{-almost everywhere.}$$

The paper by Grzaslewicz (1987) includes into the characterization of exC the case of infinite measures P_1, P_2. Moreover the infinite doubly substochastic matrices are studied.

The basic questions of the discrete marginal problem have been solved, even if there are still open problems. The validity of equivalence (4) makes the situation clearer. Combinatorial tools enable thorough investigation of the support of extremal measures as well as they may help in the existence problems.

3. The continuous case.

We may split the known results into three groups:

a) functional-analytic results
b) measure theoretic techniques based on rectangles   (6)
c) methods studying directly supports on graphs.

In the first group the review is started by the paper of Brown (1965) working with Markov operator T acting on L_∞ space associated to a σ-finite measurable space $(X_1, \mathfrak{F}, m)$. An invertible measure preserving transformation φ determines a Markov operator T_φ by the formula $T_\varphi f(x) = f(\varphi x)$. Each T_φ is an extreme point in the set M of all Markov operators. In the case of finite m, M may be identified with the set of all doubly stochastic measures on the product space $(X_1 \times X_1, \mathfrak{F} \times \mathfrak{F}, m^2)$. The main result of the paper is that M is compact in the weak operator topology and that the set Φ of operators T_φ is dense in M. The extreme d.s. measure $\mu_\varphi(AxB) = m(A \cap \varphi^{-1}B)$ is concentrated on the graph of φ. Of course by such a measure not all extreme points are represented, as shown e.g. by Ryff (1963).

A famous functional analytic characterization of extreme d.s. measures has been given independently by Douglas (1964) and Lindenstrauss (1965): μ is extreme if and only if the subspace consisting of all functions h on $X_1 \times X_2$ of the form

$$h(x,y) = f(x) + g(y), \ f, \ g \in L_1(m),$$

is norm-dense in $L_1(\mu)$. Moreover Lindenstrauss proved that each extreme d.s. measure is singular with respect to the two-dimensional Lebesgue measure m^2.

From this a common belief appeared that extreme points in the set C in (2) have mass concentrated along line segments and authors have started to attack the problem (3).

To the second group of papers belongs Brown and Shiflett (1970). They achieved the following geometrical results working on $X = I \times I$. A measurable rectangle AxB is called a full rectangle, relative to a d.s. measure μ, if the marginal measures μ^1 and μ^2 are equivalent to the one-dimensional Lebesgue measure m on A and B, respectively.
A sequence of rectangles $A_1 \times B_1, \cdots, A_{2n} \times B_{2n}$, each of which is full with respect to a d.s. measure μ, is a loop if

$$B_1 = B_2, \ A_2 = A_3, \ B_3 = B_4, \ldots, \ B_{2n-1} = B_{2n} \quad \text{and} \quad A_{2n} \subset A_1.$$

A loop is invariant if there exists a set $D_1 \subset B_1$ such that $m(D_1) > 0$ and for m-almost all $y_1 \in D_1$ the path of μ-density points

$$(x_i, y_i) \in A_i \times B_i \quad \text{with } x_{2k} = x_{2k+1} \text{ and } y_{2k-1} = y_{2k}$$

satisfies $x_1 = x_{2n}$. Under these definitions the following assertion holds: if μ is an extreme d.s. measure, then μ has no invariant loops.

A linear loop is a sequence of μ-full rectangles $A_i \times B_i$, $i=1,...,2n$, such that

$$B_1 \subset B_2, \quad A_2 \subset A_3, \quad B_3 \subset B_4, ..., B_{2n-2} \subset B_{2n-1} \text{ and } m(A_{2n} \cap A_1) > 0;$$

moreover, for any other combination of i and j, it is $m(A_1 \cap A_j) = m(B_i \cap B_j) = 0$.

A sufficient condition for an extreme measure is the following: If the d.s. measure μ has no near loops, then μ is an extreme point.

Lesche and Martignon(1980), dealing with $X = I \times I$, denote with $H(A)$ the set of all μ-density points for a given measure μ on $\mathcal{B}$ and a set $A \in \mathcal{B}$. These are points $(x,y) \in X$ for which for all $\epsilon > 0$ it is

$$\mu[A \cap ((x-\epsilon, x+\epsilon) \times (y-\epsilon, y+\epsilon))] > 0.$$

Denoting with p_x, p_y the projections of X onto the first, second coordinate, respectively, it is recursively defined

$$X_0 = p_X(H(A)), \quad Y_1 = p_y(H(X_0 \times I)) - p_y(H(A)), \quad \text{for } i > 0$$

$$X_i = p_X(H(I \times Y_i)) - X_{i-1}, \text{ for } i > 1, \quad Y_i = p_y(H(X_{i-1} \times I)) - Y_{i-1}.$$

If for all i the Lebesgue measure $\lambda(Y_i \cap p_y(H(A))) = 0$ then A is called a self-avoiding set. The theorem says that if $I \times I$ can be expressed as a countable union of closed self-avoiding rectangles, then the d.s. measure μ is extremal. Another theorem yields a necessary condition for extremity, the use of both criterions is demonstrated in examples.

Feldman conjectured that if μ_1 and μ_2 are d.s. measures with $\mu_1 << \mu_2$ (μ_1 is absolutely continuous with respect to μ_2) and if μ_2 is extreme, then $\mu_1 = \mu_2$. Among those who have considered this problem was Douglas (1964) who proved the following result: If F is a vector-lattice that is weak*-dense in $L_\infty(\mu)$, if μ is extreme in the set of d.s. measures, and if ν is a d.s. measure with $\nu << \mu$ such

that $\int_X fd\nu = \int_X fd\mu$ for all bounded f in F, then $\nu = \mu$.

Brown and Shiflett (1970) proved that Feldman's conjecture holds for the following class of extreme d.s. measures: if μ_1 and μ_2 are d.s. measures such that μ_2 has no near loops and $\mu_1 << \mu_2$, then $\mu_1 = \mu_2$.

Nice counterexamples which brought more light into the theory were constructed by Losert (1982). First he proved that a slightly stronger property than Feldman's conjecture (μ_1 need not be positive) is equivalent to the weak*-density of

$$F = \{(x,y) \rightarrow f(x) + g(y); \ f, g \in L_\infty(m)\}$$

in $L_\infty(\mu_2)$. Furthermore an extreme d.s. measure supported by the union of two graphs in X which does not fulfil the above condition is defined. Thus, Feldman's conjecture is disproved.

In the second part of the paper an example is given showing that the support of an extreme d.s. measure may be the whole unit square, i.e. the measure is not concentrated on a closed set whose m^2 measure is less than one. This measure has moreover the property that any graph of a measurable function f: I → I has m^2-measure zero. This paper shows how hard is perhaps the problem of finding a geometrical characterization of extreme d.s. measures.

Therefore it is not surprising that recent papers are devoted to the special case of extreme measures supported by some graphs of maps between X_1 and X_2, which is the third group in (6). Such measures are of interest in practical situations. Two have been mainly tackled:

a) problem of uniqueness. If in the continuous analogue of (5) F is a system of measurable graphs in X and there exists a unique measure in C_F, then it is necessarily extremal in C. There are several results giving conditions for F being a set of uniqueness.

b) problem of existence. The question whether C_F is nonempty or not is difficult to address and less results have been obtained. For example let $F=G(f) \cup G(g)$, f: $X_1 \rightarrow X_2$, g: $X_2 \rightarrow X_1$ are increasing homeomorphisms, G(.) their graphs, $X_1=X_2=I$ and s_f, s_g are the mass-spreaders of a probability measure μ supported by F, i.e.

$$s_f(x) = \mu\{(<0,x> \times I) \cap G(f)\},$$

$$s_g(x) = \mu\{(I \times <0,x>) \cap G(g)\}.$$

Then for $\mu \in C_F$ its mass-spreaders must satisfy functional equations

$$s_f(x) + s_g(g^{-1}(x)) = P_1(<0,x>) \tag{7}$$

$$s_g(x) + s_f(f^{-1}(x)) = P_2(<0,x>)$$

The characterization of s_f, s_g has been still attacked in a special situation only.

A pioneering work related to this field is Vitale and Pipkin (1976), which is formulated in terms of the class of regression functions associated to the probability distributions on the unit square with uniform marginal distributions.

Seethoff and Shiflett (1978) assume f, g : I→I, f≤g, f≠g on D⊂I. For $F=G(f) \cup G(g)$ the elements of C_F are expressed as

$$\mu_\alpha(S) = \int \alpha(x) \, \chi_S(x,f(x)) + (1-\alpha(x)) \, \chi_S(x,g(x)) \, m(dx)$$

for some measurable map α : I→I, S⊂X, χ_S the indicator function of S and m the Lebesgue measure on I. If f and g are one to one on D, then F is a set of uniqueness. If f and g are increasing homeomorphisms such that f(x)<x< g(x), x ∈ (0,1), and $\mu_\alpha \in C_F$ then m-almost everywhere α is a derivative of the function Q

$$Q(x) = f(x) - g^{-1} f(x) + fg^{-1} f(x) - g^{-1} fg^{-1} f(x) +.... \tag{8}$$

which is correctly defined and absolutely continuous on X.

Concerning the existence problem we fix a, b∈I and suppose that f and g

are bounded continuous functions on $D_1 = < 0,a)$ such that on D_1 it is $f(D_1) = <$
$0,b >$ and $g(D_1) = <b, 1>$. Moreover for the derivatives f', g' it holds
$$0 < f' \leq 1, \quad g' > 0, \quad f' + g' > 1.$$
Then f and g can be extended to X_1 in such a way that $C_F \neq \emptyset$. Some more
special cases are also investigated in the paper.

The existence results are improved in Sherwood and Taylor (1988) for the
so-called hairpin support F which is the union of graphs $G(g) \cup G(g^{-1})$ for some
increasing homeomorphism $g : I \rightarrow I$ such that $g(x) < x$ whenever $0 < x < 1$.
Finding a d.s. measure μ with support F is equivalent to finding the mass-
spreader $s_g(x)$. The equations (7) reduce now to a single equation

$$s_g(x) + s_g \circ g^{-1}(x) = x$$

whch has only one non-negative solution s_g. A hairpin supports a d.s. measure if
and only if this functional equation is satisfied for some s_g which has two
following properties:

a) $s_g(x)$ and $x - s_g(x)$ are increasing homeomorphisms from I onto $< 0, \frac{1}{2}>$

b) $s_g(x) < \frac{x}{2}$ whenever $0 < x < 1$.

Some special results on hairpins follow, e.g. if g is convex and

$$s_g(x) = \sum_{n=1}^{\infty} (-1)^{n-1} g^n(x),$$

then s_g is nondecreasing over $0 \leq x < 1$ if and only if

$$m(x) = \sum_{n=-\infty}^{\infty} (g^{2n}(x) - g^{2n+1}(x)) = \frac{1}{2}$$

for all $x \in (0,1)$.

A more general result for d.s. measures is given in Kaminski et al.(1988).
Again $F = G(f) \cup G(g)$, where f and g map I into itself and moreover

a) f and g are nondecreasing,
b) $(f \circ g)(t) < t$ and $(g \circ f)(t) < t$ whenever $0 < t < 1$.

Under these conditions F is the set of uniqueness, i.e. card $C_F \leq 1$.

4. A more general approach.

Coming back to the general setting of the marginal problem we shall assume that
X_i are Polish spaces and $\mathcal{B}_i = \mathcal{B}(X_i)$ their Borel σ-algebras, $i=1,2$. The extreme
measure of the set C in (2) (for given P_1, P_2) will be called simplicial measure. A
set $D \in \mathcal{B}$ is said to be a set of marginal uniqueness (MU-set, compare with Letac
(1966)) if

$$Q(D) = P(D) = 1, \quad Q^i = P^i, \quad i = 1,2 \quad \Rightarrow \quad Q = P.$$

A motivation for the definitions is provided by the following result (Benes, Stepan (1987)), proof of which is heavily supported by the "density theorem" of Douglas (1964).

If P is a simplicial measure then

$$P \left(\overset{\infty}{\underset{n=1}{U}} K_n \right) = 1 \qquad (9)$$

for some nondercreasing sequence of compact MU-sets $K_n \in X$.

Hence the first problem to handle is to describe the geometry of MU-sets. Towards this we consider maps $f : X_1 \rightarrow X_2$, $g : X_2 \rightarrow X_1$ denoting by Dom(f) and Dom(g) the domains of f and g, respectively.

For such a couple of maps we define $T : X_1 \rightarrow X_1$ by

$$T(x) = \begin{cases} (g \circ f)(x) & x \in D(T) = f^1(Dom(g)) \cap Dom(f) \\ x & x \in X_1 - D(T) \end{cases}$$

and call the maps f and g to be aperiodic if

$$T^n(x) = x \text{ for some } n \in N \text{ implies that } x \notin D(T)$$

(compare with Denny (1980)). Borel measurable maps f and g are said to be measure-aperiodic if any T-invariant probability measure P_1 on $\mathcal{B}_1$ is supported by $X_1 - D(T)$. If $F = G(f) \cup G(g)$ for some aperiodic (measure-aperiodic) maps f and g such that $G(f) \cap G(g) = \emptyset$, we call this decomposition of given $F \in \mathcal{B}$ an aperiodic (measure-aperiodic) decomposition of D.

Obviously, a measure-aperiodic decomposition is an aperiodic one, the contrary being true for discrete case only.

Summarizing the results concerning the discrete situation of X_1, X_2 at most countable (Letac, 1966; Denny, 1980) we get that the following statements concerning $F \subset X = X_1 \times X_2$ are equivalent (compare with (4)):

a) F is a MU-set
b) There is an aperiodic decomposition of F
c) There is no cycle in F.

Moreover, P is a simplicial measure if and only if it is supported by an MU-set $F \subset X$.

Unfortunately, this result is almost completely false in the continuous case. To verify this and some other preceeding negative statements, too, we consider the following situation: $X_1 = X_2 = I$, $f(x) = x$, $g(x) = x + \alpha \mod(I)$ for $x \in I$, $\alpha \in I$ irrational number. The ergodicity of $T = g \circ f$ implies that $F = G(f) \cup G(g)$ is an aperiodic decomposition that is not measure-aperiodic, F is not an MU-set.

Another example in Benes and Stepan (1987) shows that there exists a simplicial measure P such that $P(F_1) < 1$ for all MU-sets F_1. A set F is constructed which is not an MU-set; hence it does not admit measure-aperiodic decomposition and yet supports a simplicial measure. Moreover, there are compact MU-sets K_n such that $K_n \uparrow F$, and the implication (9) cannot be reversed.

The main result of the paper is the following theorem: Assume that $F = G(f) \cup G(g)$, $G(f) \cap G(g) = \emptyset$, where f,g are measurable mappings whose

graphs are measurable in $\mathfrak{B}$. Then F is an MU-set if and only if the mappings f
and g are measure-aperiodic.

To compare the result with the recent achievements in the theory of
extreme doubly stochastic measures referred to in Section 3 one·can see that it
concerns mainly the uniqueness problem, i.e. the problem to characterize
geometry of MU-sets. As the characterization is transformed into the problem of
the existence of invariant measure, one may use some well-known theorems
related to the problem (Halmos (1946), Roberts (1974), Pianigiani (1981),
Nadkarni (1987)) to get some negative answers.

As far as the proof is concerned the "sufficient part" is the difficult one.
For the sake of simplicity assume that Dom f=X_1, Dom g=X_2.

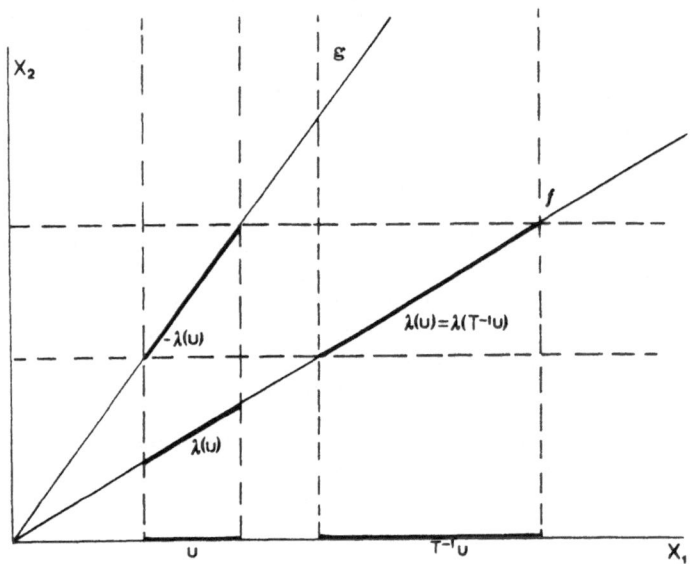

Take probability measures $P \neq Q$ supported by F such that $P^i = Q^i$, i=1,2. Denote
$m = P - Q$, $\lambda = (m|G(f))^1$ ($\lambda \neq 0$ as $P \neq Q$). As $P^i = Q^i$ for i=1,2 $m(UxX_2)=$
$= m(X_1 xg^{-1}U) = 0$, hence $\lambda(U) = \lambda(T^{-1}U)$ (see Figure) holds for any $U \in \mathfrak{B}_1$.
If $\lambda^+(X_1) > 0$, then $p = (\lambda^+(X_1))^{-1} \lambda^+$ is obviously a T-invariant probability
measure on $\mathfrak{B}_1$.

Concerning the characterization of the support of a simplicial measure
(problem (3) in Introduction) we are able to state that (X_1, X_2 Polish spaces):

Any simplicial measure is supported by a non-decreasing union of compact
MU-sets, hence (by a transfinite construction) any simplicial measure is
supported by a set which admits an aperiodic decomposition.

This result, however, is depreciated by the use of the axiom of choice. In
Section 5 of Benes and Stepan (1987) the following general problem was posed: Is
it possible to rearrange a nonmeasurable decomposition F = G(f)∪ G(g), F∈ $\mathfrak{B}$,
into a measurable one? Having received the positive answer in a very special
situation we learn from Losert (1982) that the answer is generally negative.

5. Extremal measures and Choquet theory.

The methods used in Section 4 to get a measure-aperiodic decomposition of the support of a simplicial measure may be employed in a more general context.

Consider a Polish space X, $\mathcal{B}=\mathcal{B}(X)$ being its Borel σ-algebra and denote by $\mathcal{P}(X)$ the set of all Borel probability measures defined on X. Further, consider a set A of Borel measurable real functions on X such that $1\in A$. A measure $P\in \mathcal{P}(X)$ is said to be A-simplicial if it is an extremal point in the convex set

$$C(A,P) = \{Q\in \mathcal{P}(X): A \subset L_1(Q), \int_X adP = \int_X adQ \text{ for all } a\in A\}.$$

Remark that the existence of an A-simplicial measure is guaranteed in many nontrivial situations. If follows, for example, from the representation theorem stated in the Introduction that it is the case when $A = A_1\cup A_2$ where $A_1\subset C(X)$ (the space of bounded continuous functions on X) and A_2 is at most countable set.

Further, remark that choosing $A = C(X_1)\oplus C(X_2)$ in the context of the marginal problem in Section 4, it is obvious that $P\in\mathcal{P}(X_1xX_2)$ is simplicial (in the marginal problem) if and only if it is A-simplicial.

Similarly to Section 4 it is natural to define a Borel set $K \subset X$ to be an A-set if

$$P,Q\in\mathcal{P}(K), \int_K adP= \int_K adQ, \ a\in A, \text{ implies that } P=Q.$$

Coming back again to the marginal problem it is easy to see that a set of marginal uniqueness $K \subset X_1xX_2$ is simply an A-set for $A=C(X_1) \oplus C(X_2)$. Using the theorem of Douglas (1964) it is obtained (Stepan, 1979):

If P is an A-simplicial measure then

$$P(\overset{\infty}{\underset{n=1}{\cup}} K_n) = 1 \text{ for some compact A-sets } K_1\subset K_2 \subset.... \tag{10}$$

If moreover $A \subset C(X)$ is a vector lattice then the above condition is sufficient and necessary for a $P\in\mathcal{P}(X)$ to be A-simplicial.

The result may also be used when studying the concept of simpliciality arising in Choquet theory. Assume that Z is a nonempty convex compact metrizable set in a locally convex linear topological space E. Choosing for A this time the set of all continuous affine functions defined on Z we understand that $Y\in\mathcal{B}(Z)$ is an A-set if and only if

$$P,Q\in\mathcal{P}(K), \int_Y xdP= \int_Y xdQ \text{ implies that } P=Q$$

and that a measure $P\in\mathcal{P}(Z)$ is A-simplicial if and only if it is an extremal point in

$$C(P) = \{Q\in\mathcal{P}(Z): \int_Z xdP= \int_Z xdQ\}.$$

We have denoted with

$$\int_Z x dP = b(P)$$

the Pettis integral with respect to P, called also the barycentrum of P, i.e. the integral b(P) is determined by

$$a(b(P)) = \int_Z a(x) \, P(dx), \quad a \in A.$$

It is the concept of a (regular) simplex which is the relevant one when trying to adapt (10) to our situation. Recall thet $Y \subset Z$ is a (regular) simplex if it is a compact convex set (with the compact boundary exY) such that for every $y \in Y$ there is a unique measure $P \in \mathcal{P}(exY)$ such that

$$y = \int_{exY} x dP.$$

Assertion (10) can be rewritten after some analysis in the following form (Stepan, 1977):
If $P \in \mathcal{P}(Z)$ is an A-simplicial measure then there is a sequence of regular simpleces Y_n such that

$$Y_n \subset Y_{n+1}, \quad exY_n \subset exY_{n+1} \text{ (i.e.} Y_n \text{ is a face in } Y_{n+1}), \tag{11}$$

$$P(\bigcup_{n=1}^{\infty} Y_n) = 1, \quad P(Y_n - exY_n) = 0, \, n \in N.$$

Assuming, moreover, that Z belongs to a closed hyperplane missing the origin in E it follows from (11) that:
If P is an A-simplicial measure then P(exL)=1 for some σ-compact simplex L with σ-compact boundary exL, the condition being sufficient if the minimal hyperspace over the support of P has finite dimension, see also E. Alfsen (1971), W. Szapiel (1979).
To get a necessary and sufficient condition for the A-simpliciality generally it is followed again Stepan (1977) to define a set $Y \in Z$ to be a P-simplex for some $P \in \mathcal{P}(Z)$ if

a) Y is a regular simplex and
b) the convex sets Y and b(H(Y,P)) are affinely independent, where it was denoted
$$H = H(Y,P) = \{Q \in \mathcal{P}(Z-Y) : Q \le aP \text{ for some } a \in R^+\}.$$

Recall that sets Y and bH are affinely indipendent if every point z in their convex hull can be expressed by a unique convex combination $z = \alpha y + (1-\alpha)bh$ for some $y \in Y$ and $bh \in bH$.
Now, it follows from Proposition in Stepan (1977), p.245, thet $P \in \mathcal{P}(Z)$ is A-simplicial if and only if there is a sequence of P-simpleces Y_n such that

$$Y_n \subset Y_{n+1}, \quad exY_n \subset exY_{n+1}, \quad P(Y_n - exY_n) = 0, \quad P(\cup Y_n) = 1 \tag{12}$$

Coming back to the problem to characterize simplicial measures in the marginal problem, i.e. to characterize the measures $P \in \mathcal{P}(X_1 x X_2)$ which are extremal in

$$C(P) = \{Q \in \mathcal{P}(X_1 x X_2) : \text{Marg}Q = \text{Marg}P\},$$

where Marg: $\mathcal{P}(X_1 x X_2) \to \mathcal{P}(X_1) x \mathcal{P}(X_2)$ denotes the map given by $\text{Marg}Q = (Q^1, Q^2)$, it is possible to translate the problem into the setting of A-simplicity which we have just treated.

Assume that X_1, X_2 are compact metric spaces and put as before $X = X_1 x X_2$. Denote by E_i the space of all bounded signed measures on $\mathcal{B}(X_i)$ and consider $E = E_1 x E_2$ which is a locally convex vector topological space when endowed with the product of the weak topologies on E_1 and E_2, respectively. Further consider its compact convex subset $Z = \mathcal{P}(X_1) x \mathcal{P}(X_2)$ and embed $X_1 x X_2$ into Z by the continuous injection given by

$$j(x_1, x_2) = (\epsilon_{x_1}, \epsilon_{x_2}), (x_1, x_2) \in X_1 x X_2$$

The map may be naturally extended to the continuous affine injection

$$i : \mathcal{P}(X_1 x X_2) \to \mathcal{P}(Z), \text{ where } iP = Pj^{-1}.$$

It is fairly obvious that $b(iP) = \text{Marg}P$ for $P \in \mathcal{P}(X_1 x X_2)$ and denoting the space of affine continuous functions defined on Z by A we get that

$K \subset X_1 x X_2$ is a compact MU-set if and only if

$Y = \overline{\text{conv}(jK)} \subset \mathcal{P}(X_1) x \mathcal{P}(X_2)$ is a regular simplex   (13)

(such that $jK = exY$)

and

$P \in \mathcal{P}(X_1 x X_2)$ is a simplicial measure if and only if   (14)

$iP \in \mathcal{P}(Z)$ is an A-simplicial measure.

Denoting again $H(K,P) = \{Q \in \mathcal{P}(X-K) : Q \le aP \text{ for some } a \in R^+\}$ for $K \subset X$ and $P \in \mathcal{P}(X)$ we get as a consequence of (12) the following theorem:

$P \in \mathcal{P}(X_1 x X_2)$ is a simplicial measure if and only if there is a sequence of compact MU-sets $K_1 \subset K_2 \subset ....$ such that $P(\cup K_n) = 1$ and

$\text{Marg}\mathcal{P}(K_n)$, $\text{Marg}H(K_n,P)$ are affinely independent convex sets. (15)

To see that this theorem is implied by (12) put $Y_n = \overline{\text{conv}(jK_n)}$, use (13), (14) and observe that

$$\text{Marg}\mathcal{P}(K_n) = b(\mathcal{P}(jK_n)) = Y_n$$

and

$$\text{Marg}H(K_n,P) = b(H(jK_n,iP)) = b(H(Y_n,iP)).$$

To understand (15) more clearly we denote for a $B \in \mathcal{B}(X)$ by M(B) the space of all finite Borel signed measures supported by B and put

$$M(B,P)=\{\lambda \in M(X - B) : |\lambda| \leq aP \text{ for some } a \in R^+\}$$

where P is a probability measure on X.

As (15) is equivalent to the requirement that the affine hulls of the convex sets are disjoint we get that (15) holds if and only if

$$\lambda_1 \in M(K_n), \ \lambda_2 \in M(K_n,P), \ \text{Marg } \lambda_1 = \text{Marg } \lambda_2 \text{ implies that } \lambda_1 = \lambda_2 = 0.$$

6. Multivariate problems.

In the last section some generalizations of the marginal problem to more than two dimensions are discussed. First an interesting result of Shortt (1987) is mentioned, which is still related to the theory of doubly stochastic measures, see Section 3. Remember the result by Lindenstrauss (1965) that every extreme d.s. measure is singular with respect to planar Lebesgue measure. Now let $L_1,....,L_m$ be lines through the origin in R^2 and ν any probability measure on R^2. One can consider the convex set of probabilities on R^2 whose projections onto $L_1,....,L_m$ agree with those of ν. The main theorem in Shortt (1987) says that the extreme points of this set are singular with respect to Lebesgue product measure. An example shows that there is no straightforward generalization of the theorem to the case m=∞.

Now the thesis of Machalicky (1986), which concerns the multivariate marginal problem and its extreme points, will be reviewed. Let

$$\mathcal{A}=\exp\{1,....,n\}, \ (X_i, \mathcal{B}_i) \text{ are probability spaces, } 1 \leq i \leq n.$$

$$(X,\mathcal{B}) = \overset{n}{\underset{i=1}{\otimes}} \ (X_i, \mathcal{B}_i)$$

and for

$$A \in \mathcal{A}(X_A, \mathcal{B}_A) = \underset{i \in A}{\otimes} \ (X_i, \mathcal{B}_i).$$

$D \in \mathcal{B}$ is the set of marginal uniqueness if the following holds: P,Q are probability measure on $(X,\mathcal{B})$ such that

$$\pi_A P = \pi_A Q, \ A \in \mathcal{A}, \ P(D)=Q(D)=1 \ \Rightarrow \ P=Q.$$

Here $\pi_A: X \to X_A$ is the projection and $\pi_A \lambda$ the marginal measure on $(X_A, \mathcal{B}_A)$ with respect to λ.

First some necessary conditions for sets of marginal uniqueness (MU-sets) are given. One possibility is to use the fact that slices of MU-sets are again MU-sets in the space with dimension n−1. The following theorem holds:

For n=3 let $D \in \mathcal{B}$ be an MU-set. For i≠j, $\{i,j\} \subset \{1,....,n\}$ denote

$$A_i=\{1,\cdots,n\}-\{i\}, \quad A_j=\{1,\cdots,n\}-\{j\} \quad \text{and}$$

$$X_{A_i} = \underset{k \in A_i}{\otimes} X_k, \quad X_{A_j} = \underset{k \in A_j}{\otimes} X_k.$$

Then there exist functions

$$h: X_{A_i} \to X_i \quad \text{and} \quad g: X_{A_j} \to X_j$$

such that

$$D = G(h) \cup G(g), \quad G(h) \cap G(g) = \emptyset,$$

G denoting the graph of a function. Specially for n=3 we obtain the decomposition of an MU-set into the union of two twodimensional surfaces.

Another possibility is to take A={1,....,n}, B={1,....,n} − A and denote

$$(X_0, \mathfrak{B}_0) = (X_A, \mathfrak{B}_A) \otimes (X_B, \mathfrak{B}_B).$$

If $D \in \mathfrak{B}$ is an MU-set, then $D \in \mathfrak{B}_0$ is an MU-set in $(X_0, \mathfrak{B}_0)$. The theorem says:

There exist functions $h : X_A \to X_B$, $g : X_B \to X_A$ such that

$$D = G(h) \cup G(g), \ G(h) \cap G(g) = \emptyset.$$

For n=3 the MU-set is decomposed into graphs of

$$h : X_1 \to X_2 x X_3, \ g : X_2 x X_3 \to X_1$$

or another h,g corresponding to the permutations of indeces.

Some counterexamples are available in the thesis. It is shown that the projections of MU-sets need not be MU-sets, or that the assertions of the above theorems cannot be converted. A set $D \subset R^3$ in constructed which is not a MU-set in $X_1 x X_2 x X_3$, however it is an MU-set in $X_0 x X_3$, $X_0 = X_1 x X_2$. The sufficient condition for an MU-set analogous to Benes and Stepan (1987) is not investigated.

The second part of Machalicky (1986) is devoted to the multivariate marginal problem with measures with finite support. In an algebric approach the following result is stated:

For the product of n spaces as above let

$$D \in \mathfrak{B}, \ cardD = M < \infty, \ D = \{x_k\}.$$

Let $p_k = P(x_k)$ be a probability measure with P(D)=1. Then a special matrix A of type (L,M) is defined satisfying Ap=B for some L and vector $B = (l_1,....,l_L)$. Finally D is an MU-set if and only if two conditions hold simultaneously:

a) $\sum\limits_{i=1}^{n} card \ (\pi_i \ D) \geq cardD + n-1,$

b) there exist M linearly independent rows in A.

In the geometrical approach again only necessary conditions are available. Assume

$$A_i = \{1,....,n\} - \{n\}, \quad X_{A_i} = \underset{i \in A_i}{\otimes} (X_i, \mathfrak{B}_i),$$

$D \in \mathfrak{B}$ is a MU-set, card D$<\infty$.

Then there exist functions h_{ik},

$$1 \leq i \leq n, \ 1 \leq k \leq n-1, \quad h_{ik}: X_i \to X_{A_i}$$

such that

$$D = \sum_{i=1}^{n} \sum_{k=1}^{n-1} G(h).$$

Again counterexamples to the finite marginal problem are presented in the end.

The last paper reviewed is Shortt (1985) where the problem of probability measures on the product $XxYxZ$ of three measurable spaces with prescribed marginals on XxY and YxZ has been studied. For separable spaces X,Y a Markov transition kernel $P(x,B)$ on $Xx\mathbb{B}(Y)$ is a probability measure on the Borel σ-algebra $\mathbb{B}(Y)$ for each $x\in X$ and a measurable function in x for each fixed B. If P_1 is another probability measure on X, the probability measure

$$P(AxB) = \int_A P(x,B)\ dP_1(x)$$

has marginal P_1 on X. Then $P(x,\cdot)$ is called a Morkov kernel for P over (X,P_1).

Suppose now that P and Q are probability measures on XxY and YxZ for Polish spaces with Borel σ-algebras. Let their common marginal on Y is P_1. $V(P,Q)$ is the set of all probability measures on $XxYxZ$ with marginals P and Q. According to Shortt (1982) $V(P,Q)$ is nonempty. There exist Markov kernels $P(y,\cdot)$ and $Q(y,\cdot)$ for P and Q over (Y,P_1). Any probability measure R in $V(P,Q)$ has also a Markov kernel $R(y,.)$ over (Y,P_1). Denoting with $S(M_1,M_2)$ the set of probability measures on XxZ with marginals M_1 and M_2, we call $R(y,\cdot)$ a canonical kernel for R if

$$R(y,\cdot)\in S(P(y,\cdot),\ Q(y,\cdot))$$

for each $y\in Y$. Any Markov kernel for R may be altered on a set of $y(N_1)$ with $P_1(N_1) = 0$ to obtain a canonical kernel. For a canonical kernel $R(y,.)$ for $R\in V(P,Q)$ put $E=\{y\in Y: R(y,.)$ is extreme point in $S(P(y,.), Q(y,.))\}$, E is a Borel subset of Y.

The following theorem holds:

A probability measure R in $V(P,Q)$ is an extreme point of $V(P,Q)$ if and only if for each canonical kernel $R(y,.)$ for R over (Y,P_1), $R(y,.)$ is an extreme point of $S(P(y,.), Q(y,.))$ almost surely (P_1), i.e. $P_1(E) =1$ for the above defined set E.

The second main result concerns the uniqueness problem. The following statements are equivalent:

     a) $V(P,Q)$ is a singleton set,

     b) $P_1(D(P)\cap D(Q)) = 0$.

Here

     $D(P)=\{y: P(y,.)$ is not a point mass$\}$

and

     $D(Q)=\{y: Q(y,.)$ is not a point mass$\}$.

References

Alfsen E.M. (1971) Compact Convex Sets and Boundary Integrals Springer Verlag, Berlin.

Benes V., Stepan . (1987) "The support of extremal probability measure with given Marginals". In: M.L. Puri et al. eds., Mathematical Statistics and Probability Theory, Vol.A, D. Reidel Publishing Company, 33-41.

Bican L., Stepan J. (1985) "Choquet theorem for compact measures" Statistics and Decision.

Birkhoff G. (1946) "Tres observaciones sobre el algebra lineal". Univ. nac. Tucuman Revista Ser.A, 5, 147-150.

Birkhoff G. (1948) Lattice Theory. Revised edition, New York.

Brown J.R. (1966) "Approximation theorems for Markov operators". Pacific J. of Math. 16, 13-23.

Brown J.R., Shiflett R.C. (1970) "On extreme doubly stochastic measures". Michigan Math. J. 17, 249-254.

Dall'Aglio G. (1972) "Frechet classes and compatibility of distribution functions". Symposia Mathematica, Academic Press, 9, 131-150.

Denny J. (1980) "The support of discrete extremal measures with given marginals". Michigan Math. J. 27, 59-64.

Diego A., Germani A. (1972) "Extreme measures with prescribed marginals (finite case)". J. of Combinatorial Theory, 353-366.

Douglas R.G. (1964) "On extremal measures and subspace density". Michigan Math. J . 11, 243-246.

Grzaslewicz R. (1987) "On extreme infinitely doubly stochastic matrices". Illinois J. of Math.31.

Halmos P.R. (1946) "Invariant measures". Annals of Math. 48, 735-754.

Hammersley J.M. (1961) "A short proof of the Farahat-Mirsky refinement of Birkhoff's theorem on doubly stochastic matrices". Proc. Cambridge Philos. Soc. 57, 681.

Hoffmann A.J., Wielandt H.W. (1953) "The variation of the spectrum of a normal matrix". Duke math. J. 20, 37-40.

Choquet G. (1956) "Existence et unicite des representation integrales au moyen des points extremaux dans les cones convexes". Seminaire Bourbaki, 139, 15.

Isbell J.R.: (1955) "Birkhoff's problem 111". Proc. Amer. Math. Soc. 6, 217-218.

Kaminski A., Sherwood H., Taylor M.D. (1987) "Doubly stochastic measures with mass on the graphs of two functions". Real Anal. Exchange, 13, 253-257.

Kendall D.G. (1960) "On Infinite doubly-stochastic matrices and Birkhoff's problem 111". J. London Math. Soc. 35, 81-84.

Lesche B., Martignon J. (1980) Uber die Extremalpunkte der Menge de bestochastischen Masse uber den Messraum [0,1] x [0,1]. An. Acad. Brasil Cienc. 52, 203-206.

Letac G. (1966) "Representation des mesures de probabilite sur le produit de deux espaces denombrables, de marges donnees". Ann. Inst. Fourier, 16, 497-507.

Lindestrauss J. (1965) "A remark on doubly-stochastic measures". Amer. Math. Monthly 72, 379-382.

Losert V. (1982) "Counter-examples of some conjecture about doubly stochastic measures". Pacific J. of Math. 99, 387-397.

Machalicky M. (1986) "Extremalni rozdeleni pravdepodobnosti s predepsanymi marinalnimi rozdelenimi". Diplom thesis, Charles Univ., Prague.

Marcus M. (1960) "Some properties and applications of doubly stocastics matrices". Amer. Math. Monthly, 67, 215-222.

Mirsly L. (1959) "Inequalities for certain classes of convex functions". Proc. Edinburgh math. Soc. 11, 231-235.

Mirsky L. (1963) "Results and problems in the theory of doubly stochastic matrices". Z. Wahrsch. Verw. Geb. 1, 319-334.

Mulerjee H.G. (1985) "Supports of extremal measures with given marginals". Illinois J. of Math. 29, 248-260.

Nadkarni K.R. (1987) "Descriptive Ergodic Theory". Math. Dept. of Bombay Univ., preprint.

Pianigiani G. (1981) "Existence of invariant measures for piecewise continuous transformations". Annales Poloni Math. 40, 39-45.

Rattray B.A., Peck J.E.L. (1955) "Infinite stochastic matrices". Trans. Royal Soc. Canada, Sect.III, III. Ser., 49, 55-57.

Roberts J.W. (1975) "Invariant measures in compact Hausdorff spaces". Indiana Univ. Math. J. 24, 691-718.

Seethoff T.L., Shiflett R.C. (1978) "Doubly stochastic measures with prescribed support". Z. Wahrsch. Verw. Geb. 41, 283-288.

Sherwood H., Taylor M.D. (1988) "Doubly stochastic measures with hairpin support". Probab. Theory and Related Fields 78, 617-626.

Shortt R.M. (1983) "Strassen's marginal problem in two or more dimensions". Z. Wahrsch. Verw. Geb. 64, 313-325.

Shortt R.M. (1985) "Uniqueness and extremality for a class of multiply stochastic measures". Probab. and Math. Stat. 5, 225-233.

Shortt R.M. (1987) "The singularity of extremal measures". Real Anal. Exchange, 12, 205-215.

Szapiel W. (1979) "Points extremaux dans les ensembles convexes". Bull. Acad. Polonais Sc., 22, 939-945.

Stepan J. (1977) "Simplicial measures". Memor. Vol. of J. Hajek, Academia, Praha, 239-251.

Stepan J. (1979) "Probabilty measures with given expectations". Proc. second Prague Symp. Asympt. Statist., North Holland, 315-320.

Vitale R.A., Pipkin A.C. (1976) "Conditions on the regression function when both variables are uniformly distributed". Annals Probab. 4/5, 869-873.

Weizsacker H., Winlker G. (1980) Non compact extremal integral representations: Some probabilistic aspects. Proc. of the second Padeborn meeting on functional analysis. North Holland Math. Studies, 38, 115-148.

ON EYRAUD-FARLIE-GUMBEL-MORGENSTERN RANDOM PROCESSES

STAMATIS CAMBANIS
Statistics Department
University of North Carolina
Chapel Hill, NC   27599-3260
U.S.A.

ABSTRACT. It is shown that stationary EFGM processes in continuous-time do not have the usual regularity properties and in discrete-time they cannot display the most common types of dependence! The maximal correlation of an EFGM distribution is also determined.

1. INTRODUCTION

A particularly simple class of multivariate distributions with given marginals are the Eyraud-Farlie-Gumbel-Morgenstern (EFGM) distributions [3]. Their "consistent" form makes them natural candidates for multivariate distributions of random process. The initial interest of this study was to explore the properties of EFGM random processes. Instead it was discovered that there are fundamental problems inhibiting the definition of smooth continuous-parameter stationary EFGM processes. Their dependence structure is shown to be severely limited (Proposition 3), which prevents them from enjoying the weakest possible regularity properties, such as continuity in probability (Proposition 4) and measurability (Proposition 5). These negative results are obtained under a very weak symmetry-type condition on the support of the marginal distribution; but it is conjectured that similar results are generally valid. Discrete-time stationary EFGM processes should exist but nontrivial examples have not been constructed yet, and the most common types of dependence structure cannot be exhibited by them (Proposition 6). The resulting limitations on nonstationary EFGM processes have not been explored. The conclusion of this investigation is a warning that these simple models of dependence may be inappropriate for sampled time or spatial processes. More complex and more realistic forms of bivariate dependence need to be introduced.

We also consider, in Section 3, the maximal and minimal correlation coefficient of a bivariate EFGM distribution, and we characterize those with maximal correlation coefficient equal to one (Proposition 2).

2. EFGM DISTRIBUTIONS

A bivariate EFGM distribution $H(x_1,x_2)$ with univariate marginal distributions $F_1(x_1)$ and $F_2(x_2)$ is of the form

$$H(x_1,x_2) = F_1(x_1) \, F_2(x_2) \, \{1 + \alpha \, \overline{F}_1(x_1) \, \overline{F}_2(x_2)\} \qquad (2.1)$$

where $\overline{F}(x) = 1 - F(x)$. The admissible values of the dependence coefficient α, i.e. the values for which the right hand side of (2.1) is a bivariate distribution function, form an interval $\alpha_{min} \leq \alpha \leq \alpha_{max}$. When both marginals are absolutely continuous $\alpha_{min} = -1$, $\alpha_{max} = 1$ [3], while for general (nondegenerate) marginals

$$\alpha_{min} = -\min\{[M_1 M_2]^{-1}, \, [(1-m_1)(1-m_2)]^{-1}\},$$

$$(2.2)$$

$$\alpha_{max} = \min\{[M_1(1-m_2)]^{-1}, \, [(1-m_1) \, M_2]^{-1}\},$$

where $m_k = m(F_k)$ is the infimum and $M_k = M(F_k)$ the supremum of the set

$$\{F_k(x), \, -\infty < x < \infty\} \backslash \{0, 1\} \quad [1].$$

Here we will consider the simplest among the multivariate EFGM distributions with univariate marginals $F_1(x_1), \ldots, F_n(x_n)$ introduced in [3], i.e., those of the form

$$H(x_1, \ldots, x_n) = F_1(x_1) \ldots F_n(x_n) \, \{1 + \sum_{1 \leq k < j \leq n} \alpha_{kj} \, \overline{F}_k(x_k) \, \overline{F}_j(x_j)\}. \quad (2.3)$$

The parameters of H are the n univariate marginals $F_1, \ldots, F_n$ and the $n(n-1)/2$ constants $\alpha_{k,j}$, $1 \leq k < j \leq n$, whose admissible values are determined by the 2^n inequalities:

$$1 + \sum_{1 \leq k \leq j \leq n} \epsilon_k \epsilon_j \alpha_{k,j} \geq 0 \qquad (2.4)$$

for all $\epsilon_k = -M_k$ or $1-m_k$ [1]; and when all marginals are absolutely continuous for all $\epsilon_k = \pm 1$ [3]. Multivariate distributions of the form (2.3) are uniquely determined by their bivariate marginals. Also all their marginals (of order n-1, ... ,2) are of the same type.

3. MAXIMAL CORRELATION

Here we assume the random variables X_1, X_2 have joint distribution function (2.1), finite means μ_1, μ_2 and positive and finite variances σ_1^2, σ_2^2, and we seek to determine the range of possible values of their correlation coefficient ρ, as α in (2.1) ranges over its possible values determined by (2.2). Specifially, we are interested in how close to 1 the maximal correlation can be, and how close to -1 the minimal correlation. The marginal distributions F_1, F_2 of X_1, X_2 are kept fixed and the only free parameter is the coefficient α in (2.1) which determines their dependence structure.

A useful expression of the correlation coefficient for general marginals is [1]:

$$\rho = \frac{c_1}{\sigma_1} \frac{c_2}{\sigma_2} \, \alpha \tag{3.1}$$

where

$$c = \int_{-\infty}^{\infty} F(x) \, \overline{F}(x) \, dx = - \int_{-\infty}^{\infty} x \, [1-F(x)-F(x\text{-})] \, dF(x) \tag{3.2}$$

and $F(x\text{-})$ denotes the left limit of F at x. Since $\int_{-\infty}^{\infty}[1-F(x)-F(x\text{-})] \, dF(x) = \int_{-\infty}^{\infty} d[F(x) \, \overline{F}(x)] = 0$ [1], the expression of c in (3.2) can also be written as

$$c = - \int_{-\infty}^{\infty}(x-\mu) \, [1-F(x)-F(x\text{-})] \, dF(x). \tag{3.3}$$

As the ratios c_k/σ_k in (3.1) depend only on the marginals, we first determine their largest and smallest possible values.

<u>Proposition 1.</u>   If the distribution function F(x) has jumps of size p_k and the points x_k, then

$$\frac{c^2}{\sigma^2} \leq \tfrac{1}{3} \{1 - \textstyle\sum_k p_k^3\} =: \gamma^2, \tag{3.4}$$

and equality holds if and only if F(x) is either uniform over some interval or can be obtained from such a uniform distribution by first partitioning the interval into an at most countable number of subintervals and then replacing the mass over at least one (and possibly all) of these subintervals by an equal point mass at their midpoint.

<u>Proof.</u>   Applying Schwarz's inequality we obtain from (3.3),

$$c^2 \leq \sigma^2 \int_{-\infty}^{\infty} [1-F(x)-F(x\text{-})]^2 \, dF(x) =: \sigma^2 \, \gamma^2.$$

We now evaluate γ^2. We denote by A the set of atoms $\{x_k\}$ of F, and for simplicity by F^k the Lebesgue-Stieltjes measure corresponding to the distribution function $F^k(x)$. By splitting the range of integration $\mathbf{R} = (-\infty, \infty)$ into A and $\mathbf{R}\backslash A$ we have

$$\gamma^2 = \Sigma_k \, [1 - F(x_k) - F(x_k\text{-})]^2 \, p_k + \int_{\mathbf{R}\backslash A} [1-2\,F(x)]^2 \, dF(x).$$

The second term is evaluated as follows:

$$\int_{\mathbf{R}\backslash A} [1-2F]^2 \, dF = \int_{\mathbf{R}\backslash A} [1-4F+4F^2] \, dF$$

$$= 1-F(A)- 2\{1-F^2(A)\} + \tfrac{4}{3}1-F^3(A)\}$$

$$= \tfrac{1}{3} - \Sigma_k \, p_k + 2 \, \Sigma_k \, [F^2(x_k) - F^2(x_k\text{-})] - \tfrac{4}{3} \, \Sigma_k \, [F^3(x_k) - F^3(x_k\text{-})]$$

and substituting it into the expression of γ^2 we obtain (using $F(x_k)-F(x_k\text{-}) = p_k$),

$$\gamma^2 = \tfrac{1}{3} + \Sigma_k \, p_k \, \{ \, [1 - F(x_k) - F(x_k\text{-})]^2 - 1 \, + 2 \, [F(x_k) + F(x_k\text{-})]$$

$$-\tfrac{4}{3} \, [F^2(x_k) + F(x_k) \, F(x_k\text{-}) + F^2(x_k\text{-})] \, \}$$

$$= \tfrac{1}{3} - \tfrac{1}{3} \Sigma_k \, p_k \, [F(x_k) - F(x_k\text{-})]^2 = \tfrac{1}{3} \, \{1- \Sigma_k \, p_k^3\}.$$

Equality holds in (3.4) iff equality holds in Schwarz's inequality applied to (3.3), i.e. iff for some constant r,

$$1 - F(x) - F(x\text{-}) = r(x-\mu) \quad \text{a.e. } [dF].$$

At the points of continuity of F this gives

$$F(x) = \tfrac{1}{2} \, \{1 - r(x-\mu)\} =: ax + b \quad \text{a. e. } [dF] \text{ on } \mathbf{R}\backslash A, \qquad (3.5)$$

and at the jump points $A = \{x_k\}$ it gives

$$\tfrac{1}{2} \, \{F(x_k) + F(x_k\text{-})\} = ax_k + b. \qquad (3.6)$$

Since the right hand side of (3.5) is absolutely continuous, $F(x)$ does not have a continuous singular component. Thus it has at most two components, a uniform one described by (3.5) and a discrete one described by (3.6). Their relationship as described in (3.5), (3.6) leads to the description in the statement. $\qquad \square$

Proposition 1 is a refinement of the inequality $c^2/\sigma^2 \leq 1/3$ which was obtained for absolutely continuous F in [5] and for general F in [4].

We will now consider for simplicity the case where the marginals are equal (and nondegenerate): $F_1(x) = F_2(x) = F(x)$. Then (2.2) give

$$\alpha_{min} = -[\max(M, 1-m)]^{-2}, \quad \alpha_{max} = [M(1-m)]^{-1},$$

and $-\alpha_{max} \leq \alpha_{min} < 0$ with equality if and only if $M=1-m$. And it follows from (3.1) that the possible values of ρ are $\rho_{min} \leq \rho \leq \rho_{max}$ where

$$\rho_{min} = \frac{c^2}{\sigma^2} \alpha_{min}, \quad \rho_{max} = \frac{c^2}{\sigma^2} \alpha_{max} \tag{3.7}$$

and $-\rho_{max} \leq \rho_{min} < 0$. Proposition 1 implies that among all bivariate distributions (2.1) with absolutely continuous marginals, the ones with uniform marginals have the broadest range of correlation values: $-1/3 \leq \rho \leq 1/3$. All other marginal distributions, including those described in Proposition 1 for which c/σ takes its maximal value γ, will result in a value for c^2/σ^2 smaller than 1/3. However, a possibly broader range of values for α via (2.2) may compensate to lead to a broader range of values for the correlation coefficient ρ. We now characterize the marginals for which ρ_{max} equals 1.

Proposition 2. $\rho_{max} = 1$ if and only if $F(x)$ is discrete with two atoms with mass p and 1-p, $0<p<1$, in which case

$$\rho_{min} = \begin{cases} 1 - \frac{1}{1-p}, & 0 < p \leq \frac{1}{2}, \\ 1 - \frac{1}{p}, & \frac{1}{2} \leq p < 1, \end{cases} \quad \rho_{max} = 1, \quad 0<p<1.$$

Proof.   Necessity.   Assume F is such that $\rho_{max} = 1$. Since by (2.2), $\alpha_{max} = [M(1-m)]^{-1}$, it follows by (3.4) that

$$1 = \rho_{max} = \frac{c^2}{\sigma^2} \alpha_{max} \leq \gamma^2 \alpha_{max} = \frac{1 - \sum_k p_k^3}{3M(1-m)}. \tag{3.8}$$

Therefore F must have jumps (otherwise the RHS of (3.8) would equal 1/3) and either $0<m$ or $M<1$ or both, i.e. either a first jump of size m, say p_1, or a last jump of size 1-M, say p_2, or both. Thus $0 \leq p_1, p_2$ and $0 < p_1 + p_2 \leq 1$, and (3.8) implies

$$1 \leq \frac{1-p_1^3-p_2^3-\sum_{k\geq 3}p_k^3}{3(1-p_1)(1-p_2)} \leq \frac{1-p_1^3-p_2^3}{3(1-p_1)(1-p_2)} =: A(p_1, p_2). \qquad (3.9)$$

However, it is easy to check that in the region $0 \leq p_1, p_2$ and $p_1 + p_2 \leq 1$, the function $A(p_1, p_2)$ achieves its maximum value on the segment $p_1 + p_2 = 1$ and this maximum value is $A(p_1, 1-p_1) = 1$, $0 \leq p_1 \leq 1$. It follows from (3.9) that $1 = A(p_1, p_2)$ and $p_1 + p_2 = 1$, and thus F is discrete with two atoms.

Sufficiency. Assume F is discrete with two atoms at $x_1 < x_2$, with mass p, 1-p respectively, where $0<p<1$. Then m = p = M, and as in the necessity part

$$\rho_{max} = \gamma^2 \, \alpha_{max} = \frac{1-p^3-(1-p)^3}{3 \, p(1-p)} = 1.$$

Also from (2.2) and (3.7),

$$\rho_{min} = \gamma^2 \, \alpha_{min} = - \frac{1-p^3-(1-p)^3}{3 \, [\max(p, 1-p) \,]^2} = - \frac{p(1-p)}{[\max(p, 1-p)]^2}. \qquad \square$$

It follows from Proposition 2 that when the marginal distribution F is not discrete with two atoms, then $0 < \rho_{max} < 1$. How close to or far from 1 the maximal correlation coefficient ρ_{max} is depends on F. When F is continuous, $\rho_{max} \leq 1/3$. However, there are discrete distributions with three atoms for which ρ_{max} is as close to 1 as desirable; and the same is true for mixed distributions with two atoms and an absolutely continuous component.

4. EFGM RANDOM PROCESSES

The simple and "consistent" structure of the multivariate EFGM distributions (2.3) makes it enticing to introduce EFGM random processes $\{X_t, t\in T\}$, by requiring that all their finite dimensional distributions be multivariate EFGM, i.e., that for all n and $t_1, \dots, t_n$ in T the joint distribution function of the random variables $X_{t_1}, \dots, X_{t_n}$ be of the form

$$H_{t_1, \dots, t_n} (x_1, \dots, x_n) = F_{t_1}(x_1) \dots F_{t_n}(x_n) \times$$

$$\times \{1 + \sum_{1\leq k<j\leq n} \sum \alpha(t_k,t_j) \, \overline{F}_{t_k}(x_k) \, \overline{F}_{t_j}(x_j)\} \qquad (4.1)$$

where $F_t(\cdot)$ is the marginal distribution function of X_t, the coefficients $\alpha(t,s)$, $t\neq s$ in T, are symmetric: $\alpha(t,s)=\alpha(s,t)$, and their admissible values are determined by the inequalities (2.4):

$$1 + \sum_{1\leq k<j\leq n} \sum \epsilon_k \, \epsilon_j \, \alpha(t_k,t_j) \geq 0 \qquad (4.2)$$

for all $\epsilon_k = -M_{t_k}$ or $1-m_{t_k}$, or $\epsilon_k = \pm 1$ when the marginal F_{t_k} is absolutely continuous.

Thus an EFGM process is uniquely determined by

(i)   a family of univariate distribution functions $\{F_t(\cdot), t \in T\}$, and

(ii)   a dependence coefficient function of two variables $\alpha(t,s)$, $t \neq s$ in T, which is symmetric and satisfies the inequalities (4.2) for all integers n, distinct $t_1, \ldots, t_n$ in T and the specified values of ϵ_k.

The multivariate distributions of an EFGM random process are uniquely determined by its bivariate distributions via $\alpha(t,s)$. The dependence structure of an EFGM process is thus fully determined by $\alpha(t,s)$ via (4.1) and therein lies their simplicity as well as their limitation.

A random process is stationary if its index set T is the set of integers $\{\ldots, -1, 0, 1, \ldots\}$ or the set of reals $(-\infty, \infty)$ and its finite dimensional distributions are invariant under time shifts, i.e. if the joint distributions of $X_{t_1}, \ldots, X_{t_n}$ and of $X_{t_1+h}, \ldots, X_{t_n+h}$ are equal for all choices of n, and (integer or real) h and $t_1, \ldots, t_n$. Therefore an EFGM random process is stationary if and only if its univariate distributions are all equal:

$$F_t(\cdot) = F(\cdot) \quad \text{for all t,} \tag{4.3}$$

and its dependence coefficient function $\alpha(t,s)$ depends on t, s only through their difference:

$$\alpha(t,s) = \alpha(t-s) \quad \text{for all t} \neq s. \tag{4.4}$$

Thus a stationary EFGM random process is uniquely determined by its marginal distribution function F and a dependence coefficient function $\alpha(\tau)$, $\tau \neq 0$, which is symmetric and satisfies the constraints (4.2), via (4.4), for all $\epsilon_k = 1 - m(F), -M(F)$.

While stationary or nonstationary EFGM processes are easy to introduce via (4.1), the nature of the constraints (4.2) on the function $\alpha(\tau)$ or $\alpha(t,s)$ which governs their dependence structure, requires further careful examination in order to determine which functions are admissible and what sort of processes are these EFGM random processes. The fact that even "large" values of α, i.e., near the boundary of the admissible values, may result in only limited dependence (except for the marginal in Proposition 2), raises intriguing questions concerning the dependence structure of EFGM processes.

We first show that a weak symmetry condition on the support of the marginal F of a stationary EFGM process, namely $M(F) = 1 - m(F)$, implies a severe limitation on the possible values of the dependence coefficient function $\alpha(\tau)$. This condition is satisfied by a large class of marginal distributions, including those with m=0 and M=1, such as absolutely continuous marginals. However, the discrete distributions with two values, whose maximal correlation was shown to be 1 in Proposition 2, have m=M=p and thus generally do not satisfy this condition, except when the two values are equally likely: $p = 1/2$.

<u>Proposition 3</u>. Let $\{X_t\}$ be a stationary EFGM random process with marginal distribution F and dependence function α. If $M(F)=1-m(F)$ then $|\alpha(\tau)| \leq 1/3$ α_{max} for all $\tau \neq 0$.

<u>Proof</u>. Put $x^{-1} = M(F)=1-m(F)$ (so that $x \geq 1$). We first consider the discrete time case $T = \{... , -1, 0, 1, ...\}$. Take n=2p : even and $t_1=1, ... ,t_n=n$. Then (4.2) with all signs of the ϵ_k's positive gives

$$x^2 + \Sigma_{i=1}^{2p-1} (2p-i) \, \alpha(i) \geq 0 \qquad (4.5)$$

and with the signs of $\epsilon_1, ... ,\epsilon_p$ positive and those of $\epsilon_{p+1}, ... ,\epsilon_{2p}$ negative gives

$$x^2 + \Sigma_{i=1}^{p-1} (2p-i-2i) \, \alpha(i) - \Sigma_{i=p}^{2p-1} (2p-i) \, \alpha(i) \geq 0. \qquad (4.6)$$

Summing these up we obtain

$$\tfrac{1}{2} x^2 + \Sigma_{i=1}^{p-1} (p-i) \, \alpha(i) \geq 0. \qquad (4.7)$$

Now when $p=2p_2$, i.e., $n=2^2 p_2$, it follows likewise from (4.7) and (4.6) with p replaced by p_2, that

$$\tfrac{1}{4} (\tfrac{1}{2} + 1) \, x^2 + \Sigma_{i=1}^{p_2-1} (p_2-1) \, \alpha(i) \geq 0.$$

Continuing similarly we find that when $n=2^q$,

$$b_q \, x^2 + \alpha(1) \geq 0 \qquad (4.8)$$

where the b's are defined recursively by

$$b_{p+1} = \tfrac{1}{4} (b_p + 1)$$

and clearly $b_q \downarrow \tfrac{1}{3}$ as $q \to \infty$. Since (4.8) holds for all q, it follows that

$$\tfrac{1}{3} x^2 + \alpha(1) \geq 0.$$

Similarly, with $n=2^q=2p$ we find that when the signs of the ϵ's alternate +, −, etc., (4.2) gives

$$x^2 + \sum_{i=1}^{2p-1} (-1)^i (2p-i)\, \alpha(i) \geq 0,$$

and when the signs of $\epsilon_1, \dots, \epsilon_p$ alternate $+$, $-$, etc., while those of $\epsilon_{p+1}, \dots, \epsilon_{2p}$ alternate $-$, $+$, etc., (4.2) gives

$$x^2 + \sum_{i=1}^{p-1} (-1)^i (2p-i-2i)\, \alpha(i) - \sum_{i=p}^{2p-1} (-1)^i (2p-i)\, \alpha(i) \geq 0.$$

Summing these up we obtain

$$\tfrac{1}{2} x^2 + \sum_{i=1}^{p-1} (-1)^i (p-i)\, \alpha(i) \geq 0$$

and proceeding as before we find, letting $q \rightarrow \infty$,

$$\tfrac{1}{3} x^2 - \alpha(1) \geq 0.$$

It follows that

$$|\alpha(1)| \leq \tfrac{1}{3} x^2 = \tfrac{1}{3}\ \frac{1}{M(-m)} = \tfrac{1}{3}\ \alpha_{max}.$$

To show the inequality for $\alpha(k)$, $k>1$, we consider the stationary subsequence $\{X_{kn}, n = \dots, -1, 0, 1, \dots\}$. For the continuous-time case $\{X_t, -\infty < t < \infty\}$ the result follows by considering the discrete-time stationary EFGM process $\{X_{n\tau}, n = \dots, -1, 0, 1, \dots\}$ for fixed $\tau \neq 0$.   □

A similar upper bound on the values of $|\alpha(\tau)|$ may hold for arbitrary marginals, i.e., $0 \leq m < M \leq 1$, with $1/3$ replaced by some number in $(0,1)$ depending on m and M.

The fact that, under the condition of Proposition 3, the value of the dependence coefficient of neighboring values of the stationary time series can only be a small fraction of its maximal value is a warning of possible severe lack of regularity for the EFGM process. We consider separately the continuous-time and the discrete-time cases. The results in Proposition 3 and in the following sections are infinite dimensional, i.e., they hold for infinite stationary EFGM sequences or processes, but not for multivariate EFGM distributions.

5. CONTINUOUS-TIME STATIONARY EFGM PROCESSES $\{X_t, -\infty < t < \infty\}$

Assume the marginal F satisfies $M(F) = 1 - m(F)$ and has finite second moment. Then Proposition 3 implies that for all $\tau \neq 0$,

$$|\rho(\tau)| = \frac{c^2}{\sigma^2}\, |\alpha(\tau)| \leq \tfrac{1}{3}\ \frac{c^2}{\sigma^2}\, \alpha_{max} = \tfrac{1}{3}\ \rho_{max} \leq \tfrac{1}{3},$$

so $\rho(\tau)$ cannot be continuous, since $\rho(0) = 1$. Thus <u>the EFGM process is not mean-square continuous.</u> It is not continuous in probability either, as the next result shows.

<u>Proposition 4.</u> Let $\{X_t, -\infty < t < \infty\}$ be a stationary EFGM process with nondegenerate marginal distribution F and dependence function α. If $M(F)=1-m(F)$, then the process is not continuous in probability.

<u>Proof.</u> The process is continuous in probability iff for every $\epsilon > 0$, $P\{|X_t-X_s| > \epsilon\} \to 0$ as $t-s \to 0$. But by (4.1) and [1],

$$P\{|X_t-X_s| > \epsilon\} = \iint\limits_{|x-y|>\epsilon} d^2H_{t,s}(x,y)$$

$$= \iint\limits_{|x-y|>\epsilon} dF(x)dF(y) + \alpha(t-s) \iint\limits_{|x-y|>\epsilon} B(x)B(y) \, dF(x)dF(y)$$

where $B(x) = 1-F(x)-F(x-)$. It follows that the EFGM process is continuous in probabiity iff $\alpha(\tau)$ has left and right limits at 0, $\alpha(0\pm)$, and for every $\epsilon > 0$,

$$\iint\limits_{|x-y|>\epsilon} dF(x)dF(y) + \alpha(0\pm) \iint\limits_{|x-y|>\epsilon} B(x)B(y) \, dF(x)dF(y) = 0. \quad (5.1)$$

If (5.1) holds for every $\epsilon > 0$, letting $\epsilon \to 0$ it will hold for $\epsilon=0$ as well. But denoting, as in Proposition 1 by $\{x_k\}$ the atoms of $F(x)$ and by $\{p_k\}$ their size, we have

$$\iint\limits_{x\neq y} dF(x)dF(y) = \Sigma_k \, p_k(1-p_k) = 1 - \Sigma_k \, p_k^2,$$

$$\iint\limits_{x\neq y} B(x)B(y)dF(x)dF(y) = \Sigma_k \, B(x_k)p_k \int\limits_{y\neq x_k} B(y)dF(y) = -\Sigma_k \, B^2(x_k)p_k^2,$$

since by [1], $\int_{-\infty}^{\infty} BdF = \int_{-\infty}^{\infty} d[F(1-F)] = 0$. Thus (5.1) implies

$$1 - \Sigma_k \, p_k^2 = \alpha(0\pm) \Sigma_k \, B^2(x_k)p_k^2. \quad (5.2)$$

So far the condition $M(F)=1-m(F)$ has not been used. We now show that under this condition, if the left and right limits of $\alpha(\tau)$ at zero exit, then $\alpha(0\pm) = 0$. Put as in the proof of Proposition 3, $x^{-1}=M(F)=1-m(F)$. Applying (4.2) with $t_k=k\tau$, $k=1, \dots ,n$, and $\tau>0$, we have

$$1 + \Sigma_{i=1}^{n-1} \{ \Sigma_{k=1}^{n-i} \, \epsilon_k \, \epsilon_{k+i}\} \, \alpha(i\tau) \geq 0$$

and letting $\tau \to 0$,

$$1 + \alpha(0+) \; \sum_{i=1}^{n-1} \{ \; \sum_{k=1}^{n-i} \epsilon_k \; \epsilon_{k+i} \} \geq 0.$$

Taking all signs of the ϵ_k's positive gives

$$0 \leq x^2 + \alpha(0+) \; \sum_{i=1}^{n-1} (n-i) = x^2 + \alpha(0+) \; \tfrac{1}{2} \; n(n-1), \qquad (5.3)$$

and taking the signs of the ϵ_k's alternating: $\epsilon_k = (-1)^k \; x^{-1}$, gives

$$0 \leq x^2 + \alpha(0+) \; \sum_{i=1}^{n-1} (-1)^i (n-i) = x^2 - \alpha(0+) \; [n/2], \qquad (5.4)$$

where $[n/2]$ is the integer part of $n/2$. By (5.3) and (5.4) we have

$$- \frac{2x^2}{n(n-1)} \leq \alpha(0+) \leq \frac{x^2}{[n/2]}$$

and letting $n \to \infty$ implies $\alpha(0+) = 0$.

Now getting back to (5.2) we see that continuity in probability implies

$1 - \sum_k p_k^2 = 0$, i.e., F has only one atom of size one, i.e., F is degenerate. $\qquad\square$

We finally show that the dependence function $\alpha(\tau)$ cannot be continuous in any small neighborhood of zero; otherwise the stationary EFGM process is locally independent.

<u>Proposition 5</u>.   Let $\{X_t, \; -\infty < t < \infty\}$ be a stationary EFGM process with marginal distribution F and dependence function α. If F has finite second moment and $M(F)=1-m(F)$, and if $\alpha(\tau)$ is measurable (on $(0,\infty)$) and continuous on $(0,\epsilon)$ for some $\epsilon>0$, then $\alpha(\tau) = 0$ for $0 < |\tau| < \epsilon$ and $\alpha(\tau) = 0$ for almost all $|\tau| \geq \epsilon$.

<u>Proof</u>.   From (3.1) we have for all τ,

$$\rho(\tau) = \frac{c^2}{\sigma^2} \; \alpha(\tau) \; 1(\tau \neq 0) + 1(\tau=0).$$

Since $\alpha(\tau)$ is measurable on $(0,\infty)$ and symmetric, $\rho(\tau)$ is measurable on $(-\infty,\infty)$. Thus $\rho(\tau)$ is measurable and nonnegative definite on $(-\infty,\infty)$ and therefore [2, p. 260] it can be decomposed into $\rho(\tau) = \rho_1(\tau) + \rho_2(\tau)$, $-\infty < \tau < \infty$, where ρ_1 and ρ_2 are both positive definite, ρ_1 is continuous and $\rho_2(\tau) = 0$ for a.e. $\tau \in (-\infty,\infty)$. It follows that for all $\tau \in (0,\epsilon)$, $c^2\sigma^{-2} \; \alpha(\tau) = \rho_1(\tau) + \rho_2(\tau)$, and since α and ρ_1 are continuous, so is ρ_2. Thus $\rho_2(\tau) = 0$ for $\tau \in (0,\epsilon)$ and $c^2\sigma^{-2}\alpha(\tau) = \rho_1(\tau)$. In the proof of Proposition 4 it was shown that when $M(F)=1-m(F)$ and $\alpha(0+)$ exists, it equals 0. It follows that $\rho_1(0) = 0$, and since $|\rho_1(\tau)| \leq \rho_1(0)$ for all τ, we have $\rho_1 \equiv 0$. Hence $c^2\sigma^{-2}\,\alpha(\tau) = \rho_2(\tau)$ for all $\tau \neq 0$ and thus $\alpha(\tau) = 0$ for a.e. $\tau \neq 0$. Continuity implies $\alpha(\tau) = 0$ for $\tau \in (0,\epsilon)$.   □

Proposition 5 implies that the EFGM time series is locally independent, i.e., the random variables $\{X_t, s < t < s+\epsilon\}$ are independent, for all s, and indeed each X_t is independent of every X_s with $0 < |t-s| < \epsilon$ and of almost every X_s with $|t-s| > \epsilon$.

As continuity in probability or in mean-square and measurability are minimal smoothness properties of stationary processes, the results in this section suggest that there are no good examples of stationary EFGM processes with "smooth" dependence structure. The significance of the weak symmetry condition $M(F)=1-m(F)$ on the marginal distribution remains to be explored.

6. DISCRETE-TIME STATIONARY EFGM PROCESSES $\{X_k, k=\ldots,-1,0,1,\ldots\}$

In the discrete-time case the result of Proposition 3 simply restricts the amount of dependence between successive values of the stationary sequence but does not otherwise create any pathologies or degeneracies. Still, specific examples (which should exist) elude us. We show that the most common forms of dependence, constant, exponential and m-dependence, cannot be displayed by EFGM time series. The latter case is established for m=1,2,3, but a similar result should hold for all m; however we currently have no general argument for all m.

Proposition 6. Let $\{X_k, k=\ldots,-1,0,1,\ldots\}$ be a stationary EFGM sequence with marginal distribution F satisfying $M(F)=1-m(F)$ and dependence function α.

(i)   If $\alpha(k)=a$ for all k=1,2, ... ,then a=0.

(ii)   If $\alpha(k)=\alpha_0 a^k$ for all k=1,2, ..., for some $\alpha_0>0$ and $|a|<1$, then a=0.

(iii)   If $\{X_k\}$ is m-dependent, i.e., $\alpha(k)=0$ for $|k|>m$, then $\alpha(k)=0$ for all $k\neq 0$ when m=1,2,3.

While the assumption $M(F)=1-m(F)$ simplifies considerably the analysis, it is conjectured that the results of Proposition 6 remain generally valid.

Proof. Applying (4.2) with $t_k=k$, k=1, ... ,n, we have

$$1 + \sum_{i=1}^{n-1}\left\{\sum_{k=1}^{n-i}{}^\epsilon k\,{}^\epsilon k+1\right\}\alpha(i) \leq 0. \tag{6.1}$$

We put $x^{-1}=1-m=M$.

(i)   Taking all ϵ's equal to $1-m$ we find

$$x^2 + a \sum_{i=1}^{n-1} (n-i) = x^2 + a \tfrac{1}{2}(n-1)n \geq 0,$$

and letting $n \to \infty$ we obtain $a \geq 0$. Taking now the ϵ's with alternating signs: $\epsilon_{odd} = 1-m$, $\epsilon_{even} = -M$, gives for $n = 2N+2$,

$$x^2 + a \sum_{i=1}^{n-1} (-1)^i (n-i) = x^2 - a \tfrac{n}{2} \geq 0,$$

and letting $n \to \infty$ we have $a \leq 0$. It follows that $a = 0$.

(ii) Taking all ϵ's in (6.1) equal to $1-m$ gives

$$x^2 + \sum_{i=1}^{n-1} (n-1)\alpha_0 a^i = x^2 + \frac{\alpha_0 a}{(1-a)^2} \{n(1-a) - (1-a^n)\} \geq 0,$$

and letting $n \to \infty$ implies $0 \leq \alpha_0 a/(1-a)$, i.e., $a \geq 0$. Taking the ϵ's with alternating signs and n even gives

$$0 \leq x^2 + \sum_{i=1}^{n-1} (-1)^i (n-i)\alpha_0 a^i = x^2 - \frac{\alpha_0 a}{(1+a)^2} \{n(1+a) - (1-a^n)\}$$

and letting $n \to \infty$ implies $0 \leq -\alpha_0 a/(1+a)$, i.e., $a \leq 0$. It follows that $a = 0$.

(iii) Let $\{X_k\}$ be one-dependent. Then (6.1) becomes

$$1 + \left\{ \sum_{k=1}^{n-1} \epsilon_k \epsilon_{k+1} \right\} \alpha(1) \geq 0.$$

Taking all ϵ's equal to $1-m$ gives $1 + (n-1)(1-m)\alpha(1) \geq 0$, and letting $n \to \infty$ implies $\alpha(1) \geq 0$. Taking the ϵ's with alternating signs, $\epsilon_{odd} = 1-m$, $\epsilon_{even} = M$, gives $1 - (n-1)(1-m)M\alpha(1) \geq 0$, and letting $n \to \infty$ implies $\alpha(1) \leq 0$. Hence $\alpha(1) = 0$. Note that the assumption $M = 1-m$ is not needed in this case.

Let $\{X_k\}$ be two-dependent. Then (6.1) becomes

$$1 + \left\{ \sum_{k=1}^{n-1} \epsilon_k \epsilon_{k+1} \right\} \alpha(1) + \left\{ \sum_{k=1}^{n-2} \epsilon_k \epsilon_{k+2} \right\} \alpha(2) \geq 0.$$

Taking all ϵ's with positive sign gives

$$x^2 + (n-1)\alpha(1) + (n-2)\alpha(2) \geq 0$$

and letting $n \to \infty$,

$$\alpha(1) + \alpha(2) \geq 0.$$

Taking ϵ's with alternating signs gives

$$x^2 - (n-1)\alpha(1) + (n-2)\alpha(2) \geq 0$$

and letting $n \to \infty$,

$$\alpha(2) - \alpha(1) \geq 0.$$

It follows that

$$\alpha(2) \geq 0, \quad |\alpha(1)| \leq \alpha(2).$$

Now taking n=4N and alternating the signs of successive pairs of ϵ's we obtain

$$x^2 + \alpha(1) - (n-2)\alpha(2) \geq 0,$$

which, letting $n \to \infty$, implies $\alpha(2) \leq 0$. It follows that $\alpha(2)=0$ and $\alpha(1)=0$. (It should be noted the first two inequalities hold for all values of m and M, $0 \leq m < M \leq 1$, while the third is valid for a wide range of values around M=1−m.)

Let now $\{X_k\}$ be three-dependent. Then (6.1) becomes

$$1 + \left\{\sum_{k=1}^{n-1} \epsilon_k \epsilon_{k+1}\right\}\alpha(1) + \left\{\sum_{k=1}^{n-2} \epsilon_k \epsilon_{k+2}\right\}\alpha(2) + \left\{\sum_{k=1}^{n-3} \epsilon_k \epsilon_{k+3}\right\}\alpha(3) \geq 0.$$

Taking all the ϵ's positive gives

$$x^2 + (n-1)\alpha(1) + (n-2)\alpha(2) + (n-3)\alpha(3) \geq 0$$

and letting $n \to \infty$,

$$\alpha(1) + \alpha(2) + \alpha(3) \geq 0.$$

Taking the ϵ's with alternating signs and n even, gives

$$x^2 - (n-1)\alpha(1) + (n-2)\alpha(2) - (n-3)\alpha(3) \geq 0$$

and letting $n \to \infty$,

$$-\alpha(1) + \alpha(2) - \alpha(3) \geq 0.$$

It follows that

$$\alpha(2) \geq 0, \quad |\alpha(1)+\alpha(3)| \leq \alpha(2).$$

Now taking n=4N and alternating the signs of successive pairs of ϵ's, we obtain

$$x^2 + \alpha(1) - (n-2)\alpha(2) - \alpha(3) \geq 0,$$

and letting $n \to \infty$, $\alpha(2) \leq 0$. It follows that

$$\alpha(2) = 0, \quad \alpha(1) + \alpha(3) = 0.$$

Next taking n=6N and alternating the signs of successive triples of ϵ's, we obtain

$$x^2 + \alpha(1) - (n-3)\alpha(3) \geq 0$$

and letting $n \to \infty$, $\alpha(3) \leq 0$. Finally, taking n=3N and successive blocks of one positive ϵ followed by two negative ones, we obtain

$$x^2 - \left(\frac{n}{3} - \alpha\right)\alpha(1) + (n-3)\alpha(3) \geq 0$$

and letting $n \to \infty$, $-\alpha(1) + 3\alpha(3) \geq 0$. Since $\alpha(1) + \alpha(3) = 0$, this implies $\alpha(3) \geq 0$ which along with $\alpha(3) \leq 0$ means $\alpha(3) = 0$ and $\alpha(1) = 0$. (In this case only the argument leading to $\alpha(2) = 0$ made use of the condition M=1−m, and the conclusion is valid with the same broad range of values around it as in the two-dependent case.)

7. CONCLUDING REMARKS

There seem to be no easy remedies for the problems encountered with continuous-time EFGM processes. The lack of regularity seems to be a consequence of the simple structure of the multivariate distributions (4.1). Higher dimensional coupling, via three- and higher-dimensional marginals, will not help as continuity in probability and mean-square continuity are bivariate properties. Infinite order bivariate coupling is necessary to product regular random processes, i.e., bivariate distributions with diagonal infinite series expansions of the form

$$H_{t,s}(x,y) = F_t(x) F_s(y) \left\{1 + \sum_{p=1}^{\infty} \alpha_p(t,s) A_p(F_t)(x) A_p(F_s)(y)\right\},$$

where A_p maps a distribution function into a function; or bivariate densities with analogous diagonal infinite series expansions, similar to those of the bivariate normal densities in terms of Hermite polynomials. The corresponding expansion for the multivariate distributions would be

$$H_{t_1,\dots,t_n}(x_1, \dots, x_n) = F_{t_1}(x_1) \dots F_{t_n}(x_n) \times$$

$$\times \left\{1 + \sum_{1 \leq k < j \leq n} \sum_{p=1}^{\infty} \alpha_p(t_k, t_j) A_p(F_{t_k})(x_k) A_p(F_{t_j})(x_j)\right\}.$$

There is thus a need to introduce and study multivariate distributions of such forms and the random processes they define.

ACKNOWLEDGEMENT

The research for this paper was done with the support of AFOSR Contract No. F49620-85C-0144.

REFERENCES

1. S. Cambanis. Some properties and generalizations of multivariate Eyraud-Gumbel-Morgenstern distributions. J. Multivariate Anal. 7, 1977, 551-559.

2. E. Hewitt and K. A. Ross. Abstract Harmonic Analysis, Vol. II. Springer, Berlin, 1970.

3. N. L. Johnson and S. Kotz. On some generalized Farlie-Gumbel-Morgenstern distributions. Comm. Statist. 4, 1975, 415-427.

4. G. D. Lin. Relationship between two extensions of Farlie-Gumbel-Morgenstern distributions. Ann. Inst. Statist. Math. 39, 1987, 129-140.

5. W. R. Schucany, W. C. Parr and J. E. Boyer. Correlation structure in Farlie-Gumbel-Morgenstern distributions. Biometrika 65, 1978, 650-653.

AUTHOR INDEX

SUBJECT INDEX